Books by George Gaylord Simpson

ATTENDING MARVELS (1934)

TEMPO AND MODE IN EVOLUTION (1944)

THE MEANING OF EVOLUTION (1949)

HORSES (1951)

LIFE OF THE PAST (1953)

THE MAJOR FEATURES OF EVOLUTION (1953)

LIFE: AN INTRODUCTION TO BIOLOGY. *With C. S. Pittendrigh and L. H. Tiffany* (1957)

QUANTITATIVE ZOOLOGY. *Revised edition, with Anne Roe and Richard Lewontin* (1960)

PRINCIPLES OF ANIMAL TAXONOMY (1961)

This View of Life

THE WORLD OF AN EVOLUTIONIST

George Gaylord Simpson

A Harbinger Book

HARCOURT, BRACE & WORLD, INC.
New York

ISBN 0-15-690070-x

Library of Congress Catalog Card Number: 64-14636

Printed in the United States of America

D.3.70

evolution (Biology)

There is grandeur in this view of life, with its several powers, having been originally breathed into a few forms or into one; and that, whilst this planet has gone cycling on according to the fixed law of gravity, from so simple a beginning endless forms most beautiful and most wonderful have been, and are being, evolved.

—CHARLES DARWIN,
The Origin of Species

PREFACE

THE WORLD OF AN EVOLUTIONIST is the world I live in, the one I have been exploring for a good many years now. It is also the world in which you and all of us live, but that can be true in one sense and not in another. Everyone lives in two worlds, one public and one private. The public world is the objective, material, outer world that exists around us regardless of what we know or think about it. The private world is just what we do know and think about that public world; it is the world as it seems to us, as we perceive and conceive it. This is the inner world of our individual consciousness, and in that sense it is the world in which we most truly live. It cannot possibly embrace the whole of the public world, and it cannot be quite the same for any two people. The inner and outer worlds must correspond to some extent, because sanity and life itself require private recognition of such public facts as those of gravity, sunshine, air, food, and drink. The private world may, however, lack contact with some of the aspects, even essential aspects, of the public world, and it may include concepts that are flatly false as regards the world of objective reality.

Organic evolution is one of the basic facts and characteristics of the objective world. From one point of view it is *the* basic thing about that world, because it is the process by which the universe's greatest complexities arise and systematic organization culminates. Being the process by which we ourselves came to be, it is crucial for comprehension of our place in and relationship to the objective world. Its exclusion from the subjective world therefore must seriously falsify the latter. The importance of this comprehension, attempts to reach it, some of their failures and successes, and the nature of the processes involved are among the themes of the first part of this book.

I have mentioned that the inner world cannot fully comprise the outer world for any one person or, indeed, for all of us together. It must, however, include the literally vital aspects of the objective world, and surely it should further involve such crucial but less obviously vital aspects as evolution. Therein are the secrets of specific death and survival, and our comprehension of them will prove in the long run more supremely vital than the comparatively trivial facts of individual metabolism and procreation. It is, further, important that our perception of the world be realistic, by which I mean that our ideas of the outer world should at least not flatly contradict its objective nature and at best should make the inner world a true, even though incomplete, image of the outer. To be sure, complete realism does not seem to be necessary for mere survival, not, at any rate, for such brief survival as mankind has so far managed. It is obvious that the great majority of humans throughout history have had grossly, even ridiculously, unrealistic concepts of the world. Man is, among many other things, the mistaken animal, the foolish animal. Other species doubtless have much more limited ideas about the world, but what ideas they do have are much less likely to be wrong and are never foolish. White cats do not denigrate black, and dogs do not ask Baal, Jehovah, or other Semitic gods to perform miracles for them.

Science is the means devised to minimize the discrepancies between inner and outer worlds and to pursue the ideal of a realistic concept of the universe. This functional aspect of science in permitting and promoting realism underlies all of its more extended definitions such as those discussed in Chapter 5 of this book. It is only since the rise of modern science or science strictly speaking, which is to say within about the last four hundred years, that there has been any real progress toward objectification of the individual world view. The focal point of all science is in the life sciences, as I shall argue at some length, and here the realistic approach has been even harder to achieve. An impeding factor has been the mistaken opinion that realism or objectivity requires reduction of biological phenomena to the physical level. There has also been a failure to appreciate the existence and nature of a historical factor in science, necessarily and most obviously in biological science, the key to which, evolution, is inherently historical.

We have achieved considerable knowledge of the history of life and of the processes involved in it, of the course and the causes of evolution. Such knowledge inevitably modifies our concepts of the universe around us and of ourselves. It impinges on prescientific and nonscientific philosophical and theological concepts still dominant in the subjective worlds of many of us. An adjustment is necessary for anyone reasonable enough not simply to close his eyes to this development. Most pressing and difficult is the problem of purpose—or apparent purpose—in the universe, among living things, and in our own lives. Four chapters of this book (8 through 11) are devoted to that and related problems.

Finally, it is irresistible to speculate and to extrapolate. If evolution produced first life and finally man here on earth, may it not have done so also elsewhere in the vastness of space? Is man a finality of evolution, or may he not evolve to something higher—or lower? Such questions cannot, at present, be honestly answered with much confidence. They have sometimes been answered with far more confidence than the evidence warrants and in ways contrary to the weight of the evidence. Such irresponsibility must be exposed and opposed, and this brings a serious, even somber note into what otherwise could be taken in a spirit of good clean fun.

There, in briefest preview, are some of the things about my world as an evolutionist. Most of my life has been devoted to seeking, describing, and interpreting the concrete relics of the history of life, the objective fossil record. As I learned more at that level, more abstract but more broadly significant problems at another level of interpretation increasingly demanded attention. From time to time I have been asked for reports on these broader aspects of my explorations, and I have responded with lectures, essays, or both. This book was started as a selection of such essays, collected from their widely scattered places of previous publication. In most cases, however, it was found that simple reprinting would not do. All the previously published essays have been revised, some slightly and some radically, to bring them up to date, to eliminate repetition, and to coordinate them into a unity rather than a miscellany. To round out the treatment, I have also added three chapters (4, 11, and 13) not based on previous publications. Notes at the end of the book give the antecedents of each chapter, state where, if at all, a version of it was previously published, and suggest in general terms how much revision has occurred. (A detailed statement or explanation of changes seems to me of no interest and has not been given.) Such citations and bibliographic references as seem useful are also given in the appended notes, without explicit mention in the running text.

It is my hope that this book will show that the world of an evolutionist is indeed the world of everyone. It is a further hope that it will illumine the relationships between your subjective world and the objective world, a hope that asks for consideration of my judgment of those relationships but does not demand full agreement.

G.G.S.

Los Pinavetes, La Jara, New Mexico
August 1963

CONTENTS

PART ONE

Approaches to Evolution

CHAPTER ONE

The World into Which Darwin Led Us

IT has often been said that Darwin changed the world. It has less often been made clear just what the change has been. Darwin did not—to his credit he did not—make any of the discoveries that have led to our present overwhelming physical peril. Most, although not quite all, of our technology would be the same if Darwin's work had not been done, by him or anyone else. Doubtless we would in that case still have our same traffic jams, horror movies, bubble gum, and other evidences of high civilization. The paraphernalia of civilization are, however, superficial. The influence of Darwin, or more broadly of the concept of evolution, has had effects more truly profound. It has literally led us into a different world.

How can that be? If evolution is true, it was as true before Darwin as it is today. The physical universe has not changed. But our human universes, the ones in which we really have our beings, depend at least as much on our inner perceptions as on the external, physical facts. That can be made evident by an elementary example. Suppose a stone is seen by a small boy, an artist, and a petrologist. The small boy may perceive it as something to throw, the artist as something to carve into sculpture, the petrologist as a mixture of minerals formed under certain conditions. The stone is three quite

3

different things to the three people, and yet they are seeing exactly the same thing. The stone has identical properties whatever anyone thinks about it.

In that trivial example all three conceptions of the stone, although profoundly different, are equally true. The stone can indeed be thrown, be sculptured, or be analyzed petrologically by procedures suitable to each of the three perceptions. But there are differing perceptions of objects and of our whole world that are not equally true in the same sense, which is the scientific sense of material testability. Perceptions that are not materially testable or that have been contradicted by adequate tests are not rationally valid. As they petrify into tradition and dogma they become superstitions. Perception of the truth of evolution was an enormous stride from superstition to a rational universe.

Years ago I lived for a time with a group of uncivilized Indians in South America. Their world is very different from ours: in space, a saucer a few miles across; in time, from a few years to a few generations back into a misty past; in essence, lawless, unpredictable, and haunted. Anything might happen. The Kamarakoto Indians quite believe that animals become men and men become stones; for them there is neither limitation nor reason in the flux of nature. There is also a brooding evil in their world, a sense of wrongness and fatality that they call *kanaima* and see manifested in every unusual event and object.

That level of invalid perceptions might be called the lower superstition. It is nevertheless superior in some respects to the higher superstitions celebrated weekly in every hamlet of the United States. The legendary metamorphoses of my Indian friends are grossly naive, but they do postulate a kinship through all of nature. Above all, they are not guilty of teleology. It would never occur to the Indians that the universe, so largely hostile, might have been created for their benefit.

It is quite wrong to think that uncivilized Indians are, by that token, primitive. Nevertheless, I suppose that the conceptual world of the Kamarakotos is more or less similar to that of ancient, truly primitive men. Indeed, even at the dawn of written history in the cradles of civilization, the accepted world pictures do not seem very different from that of those Indians.

The world in which modern, civilized men live has changed profoundly with increasingly rational, which is to say eventually scientific, consideration of the universe. The essential changes came first of all from the physical sciences and their forerunners. In space, the small saucer of the savage became a large disk, a globe, a planet in a solar system, which became one of many in our galaxy, which in turn became only one nebula in a cosmos containing uncounted billions of them. The astronomers have finally located us on an insignificant mote in an incomprehensible vastness—surely a world awesomely different from that in which our ancestors lived not many generations ago.

As astronomy made the universe immense, physics itself and related physical sciences made it lawful. Physical effects have physical causes, and the relationship is such that when causes are adequately known effects can be reliably predicted. We no longer live in a capricious world. We may expect the universe to deal consistently, even if not fairly, with us. If the unusual happens, we need no longer blame *kanaima* (or a whimsical god or devil) but may look confidently for an unusual or hitherto unknown physical cause. That is, perhaps, an act of faith, but it is not superstition. Unlike recourse to the supernatural, it is validated by thousands of successful searches for verifiable causes. This view depersonalizes the universe and makes it more austere, but it also makes it dependable.

To those discoveries and principles, which so greatly modified concepts of the cosmos, geology added two more of fundamental, world-changing importance: vast extension of the universe in time, and the idea of constantly lawful progression in time. Estimates of geological time have varied greatly, but even in the eighteenth century it became clear to a few that the age of the earth must be in millions of years rather than the thousands then popularly accepted from biblical exegesis. Now some geological dates are firmly established, within narrowing limits, and no competent geologist considers the earth less than 3 billion years old. (Upper estimates for the solar system range from 5 to 10 billion.) That is still only a moment in eternity, but it characterizes a world very different from one conceived as less than 6000 years old.

With dawning realization that the earth is really extremely old, in human terms of age, came the knowledge that it has changed progressively and radically but usually gradually and always in an orderly, a natural, way. The fact of change had not earlier been denied in Western science or theology—after all, the Noachian Deluge was considered a radical change. But the Deluge was believed to have had supernatural causes or concomitants that were not operative throughout the earth's history. The doctrine of geological uniformitarianism, finally established early in the nineteenth century, widened the recognized reign of natural law. The earth has changed throughout its history under the action of material forces, only, and of the *same* forces as those now visible to us and still acting on it.

The steps that I have so briefly traced reduced the sway of superstition in the conceptual world of human lives. The change was slow, it was unsteady, and it was not accepted by everyone. Even now there are nominally civilized people whose world was created in 4004 B.C. Nevertheless, by early Victorian times the physical world of a literate consensus was geologically ancient and materially lawful in its history and its current operations. Not so, however, the world of life; here the higher (or at least later) superstition was still almost unshaken. Pendulums might swing with mathematical regularity and mountains might rise and fall through millennia, but living things belonged outside the realm of material principles and secular history. If life obeyed any laws, they were supernal and not bound to the physics of inert substance. Beyond its original, divine creation, life's history was trivial. Its kinds were each as created in the beginning, changeless except for minor and obvious variations.

Perhaps the most crucial element in man's world is his conception of himself. It is here that the higher superstition offered little real advance over the lower. According to the higher superstition, man is something quite distinct from nature. He stands apart from all other creatures; his kinship is supernatural, not natural. It may, at first sight, seem anomalous that those scientists who held this view did classify man as an animal. Linnaeus, an orthodox upholder of the higher superstition, even classified *Homo* with the apes and monkeys. No blood relationship was implied. The system of nature was the pattern of creation, and it included all created things, without any mutual affinities beyond the separate placing of each in one divine plan.

Another subtler and even more deeply warping concept of the higher superstition was that the world was created for man. Other organisms had no separate purpose in the scheme of creation. Whether noxious or useful, they were to be seriously considered only in their relationship to the supreme creation, the image of God. It required considerable ingenuity to determine why a louse, for example, was created to be a companion for man, but the ingenuity was not lacking. A world made for man is no longer the inherently hostile and evil world of *kanaima,* but that again is offset in some versions of the higher superstition by the belief that man himself is inherently evil or, at least, sinful.

Those elements of the higher superstition dominated European thought before publication of *The Origin of Species,* but various studies have exhaustively demonstrated that evolutionary ideas existed and were slowly spreading among a minority of *cognoscenti* long before Darwin. Some believed that a species, although divinely and separately created, might change, and in particular might degenerate from its form in the original plan of creation. That is not a truly evolutionary view, since it does not really involve the origin of one species from another, but it does deserve to be called proevolutionary in that it recognized the fact that each separate species may change. In the eighteenth century Buffon went that far, but hardly further, in spite of some apologists who now hail him as an evolutionist.

Some eighteenth-century worthies—among them Linnaeus in his later years—did go one step further. They conceived that each of the separately created "kinds" of Genesis might later have become considerably diversified, so that the unit of separate creation might be what we now call a genus or even a family or higher group, and the species or subgroups might have arisen, or indeed evolved, since the creation. Just as the many breeds of domesticated dogs are all dogs and of common origin, so the wolves, coyotes, foxes, jackals, and other wild species might all descend from a single creation of the dog-kind. That would still admit no relationship between the dog-kind and the now likewise diversified but singly and separately created cat-kind, for example. (It is an intellectual curiosity that precisely that variation of creationist superstition has recently been seriously revived by an American who had been exposed, at least, to excellent training in zoology.)

By the end of the eighteenth century there were a few true and thoroughgoing evolutionists—Charles Darwin's grandfather Erasmus was one, as has so often been pointed out. Their number increased during the first half of the nineteenth century. Some of them even had glimmerings of Darwin's great discovery, natural selection, although (contrary to some recent historians whose aim seems to be to denigrate Darwin) none of them elucidated that principle clearly and fully.

Practically all of the ideas in *The Origin of Species* had been dimly glimpsed, at least, by someone or other before 1859. The only surprising thing about that is that so many authors have thought it worthy of special emphasis. Organization, understanding, and conviction are the main contributions of theorists like Darwin, and obviously none ever succeeded until there already existed something to organize and to understand. It is, however, less obvious why Darwin was the first evolutionist ever to carry conviction to a majority of his fellow scientists. The whole answer is more complex, but its essentials are evident in a statement later made by Thomas Henry Huxley to explain why he was an antievolutionist until he read *The Origin of Species:*

> I took my stand upon two grounds: firstly that up to that time, the evidence in favor of transmutation [evolution] was wholly insufficient; and, secondly, that no suggestion respecting the causes of the transmutation assumed, which had been made, was in any way adequate to explain the phenomena. Looking back at the state of knowledge at that time, I really do not see that any other conclusion was justifiable.

The reason why *The Origin of Species* carried conviction was that it did supply sufficient evidence of evolution and also provided an explanation of the phenomena of evolution. That twofold nature of Darwin's accomplishment has certainly been pointed out often enough, but the statement has also been criticized, and perhaps some small notice should here be given to some of the criticisms. It has, for one thing, been maintained that previous evidence *was* sufficient. It had persuaded Erasmus Darwin, Lamarck, Chambers (author of the anonymous *Vestiges of Creation*), and others, so (some critics say) it should have persuaded anyone without Charles Darwin's needing to recompile it. That conclusion is simply ridiculous. What anyone thinks *should* have happened has nothing to do with the question of historical fact. Previous evidence *did not* convince a majority of interested scientists; therefore it was insufficient for that purpose. Darwin's evidence *did* in fact convince them; therefore it was sufficient. (It may of course be recognized, as Darwin himself implied, that the way had been prepared by a changing climate of opinion and that even his evidence might have been insufficient if adduced at an earlier date.)

It has further been suggested that evolution could have been, perhaps should have been, established as a fact without requiring an explanation, and also that Darwin's explanation was not really adequate. The first proposition is debatable, certainly, and examples can be produced to support both sides. The inheritance of acquired characters was accepted by practically everyone, down to and including Darwin, even though no one had adequately explained it. Darwin himself did not like to deal with unexplained facts, and he did belatedly attempt to explain the inheritance of acquired characters. Since in this case the "facts" were not true, that particular Darwinian theory is now charitably forgotten. (Fortunately it was not really essential to his broader theory explanatory of evolution as a whole.) In any case, belief in the inheritance of acquired characters did not depend on any explanation of the supposed phenomena. (Is there perhaps a warning in the fact that the unexplained phenomena did not in truth occur?) On the other side of the argument is the modern example of extrasensory perception. A great mass of facts is claimed to demonstrate the reality of that unexplained phenomenon, and yet it is not generally accepted. It seems quite clear that it will not carry conviction unless some credible explanation is produced.

It does seem to me highly improbable that the fact of evolution would have been accepted so widely and quickly if it had been un-accompanied by an explanatory theory. Again, to question whether it *should* have been would be childish arguing with history.

The adequacy of Darwin's original explanation of evolution is also decidedly subject to debate. It was certainly an incomplete ex-planation, as Darwin was keenly aware. We now have much more extensive explanations, built in large part on Darwin's. Parts of Darwin's complex theory are also now known beyond serious doubt to have been wrong, although the more essential parts, those most stressed by Darwin, have been largely substantiated. Darwin's theory was adequate at the time in the sense of being convincing. The con-viction did not depend entirely on the truth or falsity of different parts of his explanation, which was not wholly accepted by students who nevertheless were immediately persuaded of the truth of evolu-tion. The essential point was demonstration that material causes of evolution are possible and can be investigated scientifically.

The fact—not theory—that evolution has occurred and the Darwinian theory as to how it has occurred have become so con-fused in popular opinion that the distinction must be stressed. The distinction is also particularly important for the present subject, be-cause the effects on the world in which we live have been distinct. The greatest impact no doubt has come from the fact of evolution. It must color the whole of our attitude toward life and toward our-selves, and hence our whole perceptual world. That is, however, a single step, essentially taken a hundred years ago and now a matter of simple rational acceptance or superstitious rejection. How evolu-tion occurs is much more intricate, still incompletely known, de-bated in detail, and the subject of most active investigation at pres-ent. Decision here has decidedly practical aspects and also affects our worlds even more intimately, and in even more ways, than the fact of evolution. The two will be separately considered.

The import of the fact of evolution depends on how far evolu-tion extends, and here there are two crucial points: does it extend from the inorganic into the organic, and does it extend from the lower animals to man? In *The Origin of Species* Darwin implies that life did not arise naturally from nonliving matter, for in the very last sentence he wrote, ". . . life . . . having been originally breathed by the Creator into a few forms or into one. . . ." (The words *by the Creator* were inserted in the second edition and are one of many gradual concessions made to critics of that book.) Later, however, Darwin conjectured (he did not consider this scientific) that life will be found to be a "consequence of some general law"— that is, to be a result of natural processes rather than divine intervention. He referred to this at least three times in letters unpublished until after his death, the one from which I have quoted being the last letter he ever wrote (28 March 1882 to G. C. Wallich; Darwin died three weeks later).

Until comparatively recently, many—probably most—biologists agreed with Darwin that the problem of the origin of life was not yet amenable to scientific study. Now, however, almost all biologists agree that the problem can be attacked scientifically. The consensus is that life did arise naturally from the nonliving and that even the first living things were not specially created. The conclusion has, indeed, really become inescapable, for the first steps in that process have already been repeated in several laboratories. There is concerted study from geochemical, biochemical, and microbiological approaches. At a meeting in Chicago in 1959, a highly distinguished international panel of experts was polled. All considered the experimental production of life in the laboratory imminent, and one maintained that this had already been done—his opinion was not based on a disagreement about the facts, but depended on the definition of just where, in a continuous sequence, life can be said to begin.

At the other end of the story, it was evident to evolutionists from the start that man cannot be an exception. In *The Origin of Species* Darwin deliberately avoided the issue, saying only in closing, "Light will be thrown on the origin of man and his history." Yet his adherents made no secret of the matter and at once embroiled Darwin, with themselves, in arguments about man's origin from monkeys. Twelve years later (in 1871) Darwin published *The*

Descent of Man, which makes it clear that he was indeed of that opinion. No evolutionist has since seriously questioned that man did originate by evolution. Some, notably the Wallace who shared with Darwin the discovery of natural selection, have maintained that special principles, not elsewhere operative, were involved in human origins, but that is decidedly a minority opinion about the causes or explanations, not the fact, of evolution.

It is of course also true that the precise ancestry of man is not identified in full detail and so is subject to some disagreement. That is a minor matter of no real importance for man's image of himself. No one doubts that man is a member of the order Primates along with the lemurs, tarsiers, monkeys, and apes. Few doubt that his closest living relatives are the apes. On this subject, by the way, there has been too much pussyfooting. Apologists emphasize that man cannot be a descendant of any living ape—a statement that is obvious to the verge of imbecility—and go on to state or imply that man is not really descended from an ape or monkey at all, but from an earlier common ancestor. In fact, that common ancestor would certainly be called an ape or monkey in popular speech by anyone who saw it. Since the terms *ape* and *monkey* are defined by popular usage, man's ancestors *were* apes or monkeys (or successively both). It is pusillanimous if not dishonest for an informed investigator to say otherwise.

Evolution is, then, a completely general principle of life. (I refer here, and throughout, to organic evolution. Inorganic evolution, as of the stars or the elements, is quite different in process and principle, a part of the same grand history of the universe but not an extension of evolution as here understood.) Evolution is a fully natural process, inherent in the physical properties of the universe, by which life arose in the first place and by which all living things, past or present, have since developed, divergently and progressively.

This world into which Darwin led us is certainly very different from the world of the higher superstition. In the world of Darwin man has no special status other than his definition as a distinct species of animal. He is in the fullest sense a part of nature and not apart from it. He is akin, not figuratively but literally, to every living thing, be it an ameba, a tapeworm, a flea, a seaweed, an oak tree, or a monkey—even though the degrees of relationship are different and we may feel less empathy for forty-second cousins like the tapeworms than for, comparatively speaking, brothers like the monkeys. This is togetherness and brotherhood with a vengeance, beyond the wildest dreams of copy writers or of theologians.

Moreover, since man is one of many millions of species all produced by the same grand process, it is in the highest degree improbable that anything in the world exists specifically for his benefit or ill. It is no more true that fruits, for instance, evolved for the delectation of men than that men evolved for the delectation of tigers. Every species, including our own, evolved for its own sake, so to speak. Different species are intricately interdependent, and also some are more successful than others, but there is no divine favoritism. The rational world is not teleological in the old sense. It certainly has purpose, but the purposes are not imposed from without or anticipatory of the future. They are internal to each species separately, relevant only to its functions and usually only to its present condition. Every species is unique, and it is true that man is unique in new and very special ways. Among these peculiarities, parts of the definition of *Homo sapiens,* is the fact that man does have his own purposes that relate to the future—but of man's peculiarities I have more to say below.

The heart of Darwin's explanation of how evolution occurs was natural selection. He always considered this his most important contribution, and posterity agrees with that judgment. It is true that Wallace independently but later reached almost identical views on natural selection and that several others had anticipated both Darwin and Wallace on some points. It is further true that the concept of natural selection has changed through the years since 1859 and that its major importance has occasionally been questioned. Nevertheless, natural selection was primarily Darwin's discovery, later understanding of it has developed from his, and by overwhelming consensus it is now considered the main controlling factor in most evolutionary events.

From the first edition of *The Origin of Species* Darwin expressed the opinion "that natural selection has been the main but not the exclusive means of modification." Yet in the first edition he stressed it almost to the exclusion of other factors. Summing up in the last chapter, he wrote: "Species have changed, and are still slowly changing by the preservation and accumulation of successive slight favorable variations."

That is ambiguous as to what preserves and accumulates the variations, although in context it was obvious that natural selection was supposed to do so. The ambiguity was removed by rewording in the second edition: "Species have been modified, during a long course of descent, by the preservation or the natural selection of many successive slight favorable variations."

There was considerable criticism that Darwin imputed everything, or at any rate too much, to natural selection, and he tended to retreat from so strong a stand. In the fifth edition he changed his previously flat statement by saying that modification of species occurred only "chiefly" through natural selection. In the sixth edition, 1872, the last to be fully revised, Darwin complained that he had been misrepresented, and that he had never thought modification of species due exclusively to natural selection. He made this clear, and unfortunately retreated from a stronger position, by expanding the summary of factors believed to modify species: "This has been effected chiefly through the natural selection of numerous successive, slight, favourable variations; aided in an important manner by the inherited effects of the use and disuse of parts; and in an unimportant manner, that is in relation to adaptive structures, whether past or present, by the direct action of external conditions, and by variations which seem to us in our ignorance to arise spontaneously."

That summarizes the full and final Darwinian theory, which thus recognizes four factors or causes of evolution, in sequence of importance in Darwin's opinion: (1) natural selection; (2) inheritance of acquired characters due to use or disuse of organs; (3) inheritance of acquired characters due to direct effects of the environment; (4) what we now call mutations in the broadest sense.

Darwin rejected, without even mentioning them, various dualistic, vitalistic, or otherwise nonmaterialistic theories of evolution already proposed by 1872. He accepted only factors that were believed to be strictly materialistic or naturalistic, but among those he played safe. He accepted them all, although he considered the last two unimportant as explanations of adaptation. Later in the nineteenth century there was an interesting parceling out of Darwin's four factors into three distinct theories, each emphasizing one or two of those factors at the expense of the others.

One school took the attitude of which Darwin had, as he felt, been falsely accused. They emphasized Darwin's first factor, natural selection, and flatly rejected almost any others, explicitly the inheritance of acquired characters, whether acquired from habit or from environmental influence. Their theory, more Darwinian than Darwin's, came to be called flatly Darwinism or, more specifically, Neo-Darwinism.

A second school of theory accepted and emphasized the inheritance of acquired characters, Darwin's second and third factors, and minimized without necessarily wholly rejecting the influence of natural selection and of mutation. That theory is now usually called Lamarckian or Neo-Lamarckian, but the designations are misleading. Neo-Lamarckism rejects the very heart and basis of Lamarck's personal theory, which was an idealistic and vitalistic view of continuous climbing of a "ladder of nature," from simple to complex beings. Neo-Lamarckism also stresses a factor that Lamarck rejected: inheritance of direct effects of the environment. Neo-Lamarckism is more Darwinian than Lamarckian and is, indeed, about as Darwinian as Neo-Darwinism. It emphasizes Darwin's second and third factors rather than his first one, but it does not wholly reject any Darwinian factor, and it includes nothing that was not explicitly accepted by Darwin. Lamarck's own theory, so different from Neo-Lamarckism, is discussed in Chapter 3.

The third theory here in question emphasized Darwin's fourth factor, his "variations which seem to us in our ignorance to arise spontaneously," now called mutations. This was not, however, a development of parts of Darwinian and to some extent pre-Darwinian theory, as both Neo-Darwinism and Neo-Lamarckism were. To

Darwin, mutation (not yet under that name) was only one way, and the least important way, in which materials for evolution arose. The mutationists were striking out along quite new lines, developing modern genetics and rediscovering Mendelism. The extreme mutationists, notably De Vries, held that mutations were the *only* way in which significant evolutionary change occurs. They reduced natural selection to the minor and negative role of eliminating mutants so grossly malformed as to be unviable. They agreed with the Neo-Darwinians in denying the reality of the inheritance of acquired characters.

Those three theories, tagged as Neo-Darwinism, Neo-Lamarckism, and mutationism, seemed in the early twentieth century to be the principal if not the only alternatives as naturalistic explanations of evolution. They have in common the fact that they are naturalistic. That is, they hold that evolution is a consequence of the material, physical properties of the universe and that it is explicable without postulating any immediate nonphysical, non-natural influences. Beyond that quite basic philosophical point, the three theories do lead to three different views of the world and of man's nature and potentialities.

In the Neo-Darwinian view, the crucial point in evolutionary change is the comparative success of genetic variants in producing offspring. Given a store of varying genetic materials within a population, natural selection usually tends to produce and to increase genetic combinations that are likely to ensure survival and continued reproductive success for the group as a whole. Genetic variation in itself is not considered adaptive in origin, and it is not *directly* influenced by any needs, desires, or activities of individuals in the population. Yet genetic change through the generations is decidedly nonrandom, as a rule, and tends to be adaptive for the population. To that extent, the Neo-Darwinian theory is still accepted by a majority of biologists today. It has not been rejected but only modified by being integrated into a synthesis that is both broader and deeper. The Neo-Darwinian world view originally stressed individual survival, especially competitive success. The later synthesis has involved considerable modification of that emphasis.

The Neo-Lamarckians give prime importance to exactly those factors that were minimized by the Neo-Darwinians: the needs, desires, and activities of individuals. Those factors, together with the modifying influences of soil, climate, food, and other environmental features, are supposed to lead directly to evolutionary change. Thus, genetical modification is supposed to be adaptive in its very essence. The entire process is oriented by the direct, unmediated reactions of individual organisms to their environments. The simplicity of this view is appealing, and it also has a special emotional attraction. It suggests that personal accomplishment counts not only in one's own lifetime but also in posterity and in the eventual evolution of the human species. Improvement in physique by exercise, diet, and so on, may lead to stronger descendants, and education may lead to more intelligent ones. A world in which that was true would on the whole be a pleasant one, and also one in which human progress would be comparatively easy to control. Undoubtedly it is that appeal and its political implications that have made a form of Neo-Lamarckism popular among the rulers of the Soviet Union. As I have already pointed out, there is justification for not labeling this theory with the name of Lamarck. The Russians variously call it "Soviet creative Darwinism" (as opposed to capitalistic and unacceptable Neo-Darwinism), "Michurinism," or "Lysenkoism." Although it is improbable that any of the really able Russian biologists fully accepts that theory in private, it is publicly approved Communist dogma.

The only trouble with Neo-Lamarckism in any of its various seductive guises is that it is not true. Now that we understand the mechanism of inheritance, which Darwin could not know, it is certain that acquired characters cannot possibly be inherited in the way demanded by this theory, and that is that.

The extreme mutationist world view is very different from either the Neo-Darwinian or the Neo-Lamarckian. In it evolution is dominated by chance. Change within species or from one species to another is believed to be not only initiated but also carried through by a mutation or, eventually, a sequence of mutations. Mutations certainly have definite physical causes, even though these are unknown in most specific instances, and they have determinate effects. They arise, however, by chance, and their effects are random in the sense that the cause of a mutation has no evident relationship to the nature of the result and that the effects are unoriented with respect to usefulness or adaptation in the organism. The same cause, such as radiation, may result in any and all kinds of mutations, none producing changes adaptively related to the original radiation. Furthermore, if, for instance, animals are in a situation where increase in size would be adaptive, mutations for larger size do not thereby become either more or less frequent. Mutants are in these senses random or accidental. If there just happens to be some niche into which they can fit, they survive, and a step in evolution has occurred. A recent form of the theory calls such lucky mutants "hopeful monsters." If the mutant does not happen to fit anywhere, it dies, and that is all. Evolution in the mutationist world is not merely aimless but also directionless.

That mutations occur and are random in the stated sense of that word are facts established by innumerable observations. Mutationism, unlike Neo-Lamarckism, rests on a basis of real phenomena. Nevertheless, the further deductions drawn by the original and the extreme mutationists are flatly contradicted by other phenomena, notably those of adaptation. The origin of such an organ as an eye, for example, entirely at random seems almost infinitely improbable. Added to such considerations are many paleontological examples showing evolution occurring through millions of years not fitfully and haphazardly but in a perfectly definite and manifestly adaptive way. The theory that the direction of evolution is fully controlled by mutation simply cannot be true.

Adaptation and the apparent purposefulness of evolution are basic problems that a successful theory *must* solve. The rising science of genetics early in this century not only failed to solve the problem but also made it appear insuperably difficult. That explains why almost no students of other disciplines were inclined to accept mutationism, and why Neo-Lamarckism, an elegant but as we now know incorrect solution, hung on for so long. It also was one of several reasons for the continued popularity of non-naturalistic theories, to which I allude below.

The way out of the dilemma seems simple now that it has been found. Mutationism is not an alternative to Neo-Darwinism but a supplement to it. If mutation is the source of new variation and yet is substantially nonadaptive, and if the actual course of evolution is to a large extent adaptive, then some additional factor or process must frequently intervene between the occurrence of mutations and the incorporation of some of them into evolving populations. The intervening process must be literally selective, because it must tend (not necessarily with full efficiency) to weed out disadvantageous mutations and genetical combinations and to multiply those that are advantageous in existing circumstances. Natural selection is just such a process, and the principal modern theory of evolution, although it contains much besides, is in large part a synthesis of selection theory and mutation theory.

Evolution is an extremely complex process, and we are here interested mainly in the effects of the concept on our world rather than in the process for its own sake. For that purpose I must, however, briefly note the main elements of the process now known. (Further details will be given in later chapters.) Genetic systems, governing heredity in each individual case, are composed of genes and chromosomes, discrete but complexly interacting units at different levels of size and complexity. The genes themselves, their organized associations in chromosomes, and whole sets of chromosomes have a large degree of stability as units, but all the kinds of units are shuffled and combined in various ways by the sexual processes of reproduction in most organisms. Thus, a considerable amount of variation is maintained, and, so to speak, genetic experimentation occurs in all natural populations. Mutations, in the broadest sense, affecting individual genes, chromosomes, or sets of chromosomes, introduce wholly new variation, which is fed into the processes of recombination.

Populations of similar animals, usually interbreeding among themselves and definable as species, have genetic pools, characterized by the total of genetic units in the included individuals and the distribution of combinations of those units through the population. Evolutionary change involves changes in the genetic pool, in kinds of included units, in frequencies of them, and in kinds and frequencies of combinations of them. Recombination alone does not tend to change the genetic pool. Only four processes are known to do so: mutation, fluctuation in genetic frequencies (what are known statistically as "sampling errors"), inflow of genes from other populations, and differential reproduction. The first three of those processes are not oriented toward adaptation. They are in that sense essentially random, and are usually inadaptive, although they may rarely and coincidentally be adaptive. By "differential reproduction" is meant the consistent production of more offspring, on an average, by individuals with certain genetic characteristics than by those without those particular characteristics. That is the modern understanding of natural selection, including but broader than the Darwinian or Neo-Darwinian concept, which emphasized mortality and survival more than reproduction. Natural selection in the Darwinian sense and still more in this expanded sense is nonrandom, and its trend is adaptive. It also tends, not always with complete success, to counteract the random effects of mutation and sampling error.

Evolutionary processes are tremendously more complicated in detail than this bald outline suggests. The point of the outline is that here is a mechanism, involving only materials and processes known beyond a doubt to occur in nature, capable (as one of its proponents has said) of generating just the degree of improbability evident in the phenomena of evolution.

Further information pertinent to our theme is provided by paleontology, the actual record of events in the history of life. Observation and experimentation with living organisms can extend over a few years, at most. There is always a possibility that processes there evident worked out differently over spans of millions of years, or that the actual history involved principles undetectable in shorter periods of time. There is admittedly some difference of opinion, but I think it fair to say that there is now a consensus for the view that the fossil record is fully consistent with the modern synthetic theory of evolution and that it neither requires nor suggests any alternative explanation.

There is one thing demonstrated by the fossil record that is decidedly pertinent here and that probably would never have been inferred from study of living organisms. Throughout the whole history of life most species have become extinct, without issue. The statistically usual outcome of evolution is not, then, the progressive appearance of higher forms but simply obliteration. There has, indeed, been progression and even (still more rarely) progress, but this has been in the comparatively few, exceptional lines of descent. The adaptive mechanism of natural selection has guaranteed that some lineages would win, that the world would indeed be filled and kept filled with adapted organisms, but just as inexorably it has insured that most lineages would lose. It has, moreover, had the result that even the winners, the lineages that have survived so far, have not necessarily been progressive, from a human point of view at least. The primitive ameba has remained adapted, hence has survived, while the lordly dinosaurs lost adaptation and therefore life. The degenerate tapeworm is to all appearances as well adapted as the—we like to think—progressive man.

The theory just outlined obviously does not yet answer all questions or plumb all mysteries, even when the details here omitted are taken into consideration. It casts no light on the ultimate mystery— the origin of the universe and the source of the laws or physical properties of matter, energy, space, and time. Nevertheless, once those properties are given, the theory demonstrates that the whole evolution of life could well have ensued, and probably did ensue, automatically, as a natural consequence of the immanent laws and successive configurations of the material cosmos. There is no need, at least, to postulate any non-natural or metaphysical intervention in the course of evolution.

That conclusion has been questioned or opposed not only by many philosophers and theologians but also by a comparatively small number of scientists. The alternatives occasionally supported by scientists or scientific philosophers, and therefore pertinent here, comprise many shadings and variations of opinion, but most of them can be placed in the rubrics of vitalism and finalism.

The vitalists maintain that life is an essence or principle in itself, absent in nonliving matter and not reducible to the interaction of fully material factors. They usually point to a directedness or apparent purposefulness in the development and activities of living things and conclude that the vital, nonmaterial essence within them is a controlling influence in evolution. The finalists maintain that the evolutionary history of life has a preordained over-all pattern which, at least until the appearance of man, was purposefully directed toward a future goal or end. There is no absolute logical necessity that vitalism and finalism should go together, but the ideas are related if only because both are to some degree non-naturalistic and, in that sense, nonmaterialistic. More often than not, vitalists are finalists and finalists are vitalists.

Darwin's legacy in this respect was somewhat but not altogether negative. He did not discuss these issues explicitly and in plain terms. From the whole body of his work, and perhaps more particularly from notes and letters not written for publication, it is clear enough that he felt an antipathy for these philosophical approaches. The very fact that he did not specifically go into these problems amounts to a tacit but positive stand that metaphysical postulates are not necessary for a scientific explanation of evolution.

To that extent it is quite true, as has been so often said, especially by his enemies, that Darwin was a materialist. "Materialist" has become a highly ambiguous word and in some circles a dirty one. It is better here to use the word "naturalist," in the proper philosophical sense of a scientific inquirer who eschews recourse to the supernatural. Such an inquirer does not deny the possible existence of the supernatural but only excludes it from attempts at scientific explanation. Almost all scientists agree that such exclusion is pragmatically justified and indeed necessary. Appeal to the unknown or to the scientifically untestable always stultifies the progress of science, because it stops the search for material explanations that *are* scientifically testable—which, as a matter of experience, have generally been forthcoming when the search has been continued.

Most scientific evolutionists since Darwin have followed his lead in this matter and have continued to seek material, natural explanations of evolution without necessarily taking any overt stand on vitalism or finalism. To the extent that vitalism and finalism are nontestable, that attitude is justified, and the scientist, as scientist, has no right to go further than to repeat the classic remark that he has no need of that hypothesis. However, I do not see how the matter can in all candor be dropped at that point even by the least philosophical of evolutionists, for there are repeated claims by vitalists and finalists that their views *are* testable and that there *is* need for that hypothesis.

Those matters are further discussed later in this book. Here I shall only rather flatly state conclusions. These conclusions are not accepted by all evolutionists, but I think it safe to say that they are by most. The sort of testable evidence that would suggest vitalism or finalism would be the steady progression of life, and of each of its evolving lineages, toward a final and transcendentally worthy goal. That is not, in fact, what the known record of life's history shows. There is no clear over-all progression. Organisms diversify into literally millions of species, then the vast majority of those species perish and other millions take their places for an eon until they, too, are replaced. If that is a foreordained plan, it is an oddly ineffective one. Single lineages, when they can be followed for long, often do show rather steady change, but not indefinitely. They become extinct, or, if they survive, the directions and rates of their evolution change. They evolve exactly as if they were adapting as best they could to a changing world, and not at all as if they were moving toward a set goal. As for the directedness that does indeed characterize vital processes, it is amply explicable by natural selection without requiring any less mundane cause.

That sort of evidence, with much else in detail, convinces me, at least, that the hypotheses of vitalism and finalism are not necessary. Everything proceeds as if they were nonexistent. That does not prove that they are untrue, but it makes their positive adoption unjustified.

Vitalism and finalism have one other aspect that has no particular scientific bearing but that does require mention. They are sometimes advanced with the avowed hope of retaining something from the world of superstition. Vitalism then pretends to find a place in nature for the supernatural. Finalism tries to bring in by the back door the teleology that Darwin swept out the front door. (See Chapter 10.)

Let me summarize and conclude as to this world into which Darwin led us. In it man and all other living things have evolved, ultimately from the nonliving, in accordance with entirely natural, material processes. In part that evolution has been random in the sense of lacking adaptive orientation. As a rule, however, it has been oriented or directed toward achieving and maintaining adaptive relationships between populations of organisms and their whole environments. Nevertheless, this blind, amoral process has not guaranteed indefinite maintenance of adaptation for any given lineage of populations. On the contrary, it usually leads to eventual extinction and a repeopling of the world by the newly divergent offspring of a minority of earlier successful lineages. The mechanism of orientation, the nonrandom element in this extraordinarily complex history, has been natural selection, which is now understood as differential reproduction.

Man is one of the millions of results of this material process. He is another species of animal, but not just another animal. He is unique in peculiar and extraordinarily significant ways. He is probably the most self-conscious of organisms, and quite surely the only one that is aware of his own origins, of his own biological nature. He has developed symbolization to a unique degree and is the only organism with true language. This makes him also the only animal who can store knowledge beyond individual capacity and pass it on beyond individual memory. He is by far the most adaptable of all organisms because he has developed culture as a biological adaptation. Now his culture evolves not distinct from and not in replacement of but in addition to biological evolution, which also continues.

Concomitant with these developments is the fact that man has unique moral qualities. The evolutionary process is not moral—the word is simply irrelevant in that connection—but it has finally produced a moral animal. Conspicuous among his moral attributes is a sense of responsibility, which is probably felt in some way and to some degree by every normal human being. There has been disagreement and indeed confusion through the ages regarding to whom and for what man is responsible. The lower and the higher superstitions have produced their several answers. In the post-Darwinian world another answer seems fairly clear: man is responsible to himself and for himself. "Himself" here means the whole human species, not only the individual and certainly not just those of a certain color of hair or cast of features.

The fact that man knows that he evolves entails the possibility that he can do something to influence his own biological destiny. The fact that uncontrolled evolution often leads to degeneration and usually to extinction makes it highly advisable that man take a hand in determining his own future evolution. If man proceeds on the wrong evolutionary assumptions—for instance, on those of Neo-Lamarckism or Michurinism—whatever he does is sure to be wrong. If he proceeds on the right assumptions, what he does may still be wrong, but at least it has a chance of being right.

A world in which man must rely on himself, in which he is not the darling of the gods but only another, albeit extraordinary, aspect of nature, is by no means congenial to the immature or the wishful thinkers. That is plainly a major reason why even now, a hundred years after *The Origin of Species,* most people have not really entered the world into which Darwin led—alas!—only a minority of us. Life may conceivably be happier for some people in the older worlds of superstition. It is possible that some children are made happy by a belief in Santa Claus, but adults should prefer to live in a world of reality and reason.

Perhaps I should end this chapter on that note of mere preference, but it is impossible to do so. It is a characteristic of this world to which Darwin opened the door that unless *most* of us do enter it and live maturely and rationally in it, the future of mankind is dim, indeed—if there is any future.

CHAPTER TWO

One Hundred Years Without
Darwin Are Enough

SUPPOSE that the most fundamental and general principle of a science had been known for over a century and had long since become a main basis for understanding and research by scientists in that field. You would surely assume that the principle would be taken as a matter of course by everyone with even a nodding acquaintance with the science. It would obviously be taught everywhere as basic to the science at any level of education. If you think that about biology, however, you are wrong. Evolution is such a principle in biology. Although almost everyone has heard of it, most Americans have only the scantest and most distorted idea of its real nature and significance. I know of no poll, but I suspect that a majority doubt, disbelieve, or violently oppose its clear truth without a hearing and on no basis more rational than prejudice, dogma, or superstition. Many school and not a few college teachers either share that irrationality or evade teaching the truth of evolution from other motives. That is a main reason why, as I said in the last chapter, only a minority of us have fully entered the world into which Darwin led us.

This irrational prejudice is a problem, and a very serious one, for our educational system and for the whole dream of developing the enlightened citizenry on which the ideal of democracy depends. It is not enough, then, simply to state, as I have, that everyone should enter the world into which Darwin led us. Some more personal and practical thought must be given to why everyone should enter it, why they have not, and what can be done about it. There are deep and tangled roots that cannot be followed in one short chapter, but I shall here attempt a superficial examination, at least, centered on the educational system where much of the impediment and the greater part of the hope are inherent.

Let me begin with some personal reminiscences. I want to talk especially about high schools. It still seems to me that primary schools are the places for simple routine, learning the indispensables of reading, writing, arithmetic, and association with one's peers in a disciplined situation. Colleges should be the places for the deepening of special intellectual interests, for their broader integration, and for the laying of a basis for some complex vocations. That leaves high schools as the places most appropriate for encountering special interests and for starting some intellectual orientation.

It is, I am sure, already evident from these remarks that I have never taught in a high school. Nevertheless I have had many contacts with such teaching in three different capacities and over a period of some forty-five years. First, of course, I was on the receiving end of high school education in one of the good public school systems (by the then standards) of the 1910's. I became fascinated by literature because I had teachers who were fascinated by it. I developed a dislike of history because I had teachers who thought events were things that occurred on dates and that the dates were what one should learn. After a bit of a struggle, I achieved a sound routine knowledge of mathematics at the intermediate levels. As for science, that was limited to one course called "physics," which, as far as my memory goes, consisted of measuring things (lengths, weights, times, temperatures) and making the measurements agree with the book. I learned, and later had to unlearn in order to become a scientist myself, that science is simply measurement and the answers are in print.

Nothing I then learned had any bearing at all on the big and real questions. Who am I? What am I doing here? What is the world? What is my relationship to it? Not that I did or could specify those questions at that time. I only felt them as an unformulated dissatisfaction, a sense of fatal incomprehension of my own being. If any of my teachers dreamt of formulating such questions for me, they never dared to live their dream. I believe that all adolescents, the bright equally with the dull, go through such a phase of incoherent self-questioning and disorientation, and that they still rarely receive what help and truth could be given them. They usually simply bury the questions and try to forget them, or they settle for answers that are palpably false.

My next serious contact with intermediate education was some twenty years later when my children in their turn attended high schools in another good public school system. I found that some progress had been made, but not much. A few exceptional teachers did point out that what the student determined and checked for himself could be true even if it was not in the book. All the students now had some contact with the life sciences, but the level—or at least the emphasis—was on such questions as why you should brush your teeth and why you should not drink alcohol. High school biology did then have strong personal reference. It was student-oriented, but it was most decidedly not well calculated to orient the student.

Since my own children left high school, my contacts have been through talking to groups of teachers about my field of science, especially organic evolution. Invitations to speak have always expressed the same purpose: to bring the teachers' knowledge up to date by telling them about recent research and the present status of theory among the professional researchers. I have taken that request literally and have always tried to fulfill that purpose, at least, on such occasions. Nevertheless, I found on the very first occasion and have further observed on all those since then that there is a great deal more here than is apparent at first glance. Without exception, there has always been something seriously wrong beneath the surface of all these talks, conferences, seminars, and institutes with high school teachers. It has not always been quite the same difficulty, but it has always presaged a failure or, at best, a severe limitation of the intended communication from scientist to student via teacher.

One aspect, I think not the most common one, of that failure was expressed to me by an officer of an association of high school biology teachers more or less in these words:

> As individuals, we have all learned a great deal from your exposition of modern evolutionary theory. It also helps our status in the school system to demonstrate, by attendance at these meetings, that we are keeping on our toes. But, frankly, not a thing that you have said is going to be passed on to a single student in this city. We have found that it simply is not worth while to do more than go through the barest motions demanded by the curriculum. Our students don't want more, and only one in a hundred really wants that much. We shove them on as best we can, and then we are rid of that particular batch. Our only real function is that of custodians, to keep the kids locked up and off the streets for a certain number of hours each day.

That was several years ago. Perhaps things are no longer so bad in that school system, and probably they never were quite so bad in most systems. I gather that it is no longer universally condemned as undemocratic to give separate and special opportunities to selected students who can learn and want to or who can be induced to. I, at least, can be less concerned about what to do with the other students and for their unfortunate teachers.

One of the phenomenal developments of recent years has been the great development of institutes and conferences for high school teachers, selected from all over the United States and with expenses and a reasonable salary for attending paid from federal funds. That has, by the way, raised a serious problem that is only indirectly germane to my present subject but that teachers should know about. One purpose of these activities is to bring teachers into contact with the research workers in their subjects. The institutes and conferences are now so numerous and so prolonged that success in that aim demands many thousands of man-hours that would otherwise be available for research itself rather than for talking about it. In the usual case of a university professor with time already divided between teaching and research (with, as a rule, some administration thrown in), supplying all, many or in some instances any of the high school teachers' needs means that his research must slow down or even grind to a halt. Thus, our schizophrenic government agencies are with one hand giving large sums to promote research and with the other giving likewise large sums which, in effect, impede or prevent research. Both activities are very much worth while, but some way must surely be found to mitigate the conflict between them.

What I have just said explains why I have so far taken part in only a few of the recent national institutes and conferences for high school teachers. Those at which I have spoken have been truly enjoyable and have, I think, had a reasonable measure of success in the aims of bringing research and teaching into useful relationship. Nevertheless, each one has also brought out failures of communication and emphasized the difficulty of somehow getting through to the students what they should and, in a workable modern culture, *must* know. As revealed at this level of researcher-teacher contact, the most serious of these impediments are three:

1. The knowledge of some teachers cannot, by the means here provided, be brought up to date because the teacher does not have enough knowledge, even outdated knowledge, to begin with.

2. Some teachers are quite willing to listen (they are being paid to, after all), but they are not at all willing to learn. As regards my subject, evolution, a significant minority of them simply do not believe a word of it and automatically close their minds when the subject is named.

3. In a large minority of instances—indeed it may not be a minority—the teachers themselves accept what is reasonably presented to them but still do not expect to incorporate it into their teaching because of the attitudes and power of school officials, school boards, parents, and tax-appropriating bodies.

All three of those points are frequently involved in a vicious circle. The teachers are parts of the system that produces inadequate preparation, personal bias, and community prejudice. They cannot reasonably be expected to correct defects of which they are themselves both causes and effects. It is a possible function of the institutes and similar efforts toward teacher education to break that vicious circle. That could perhaps be their most significant contribution, but the degree of success so far is open to question. The teacher who has been trained in a school with substandard staff and a watchful antiscientific board and who has then gone to teach in just such another school is not going to start giving an unbiased and modern course in biology merely because he has listened to a few scientists whom he (or she) is not prepared either to understand or to respect.

Still, that apparently quite hopeless situation is an extreme. Most teachers must suffer to some degree from one, two, or all three of these impediments, but usually not to a degree that precludes improvement. The institutes also turn up a heartening number of teachers who are surprised but receptive when they learn that research biologists and whole scholarly communities take evolution as an established fact, the fundamental fact of life, and who then are eager to learn more. An antievolutionary community cannot be directly affected by that contact, but a change in attitude can be initiated and the vicious circle finally broken if such a teacher is able to pass on something of this aspect of biology to new generations of students. That is by no means easy, however, nor is success assured even to the most convinced, determined, and tactful teacher.

The pressures of some communities happen to be particularly strong in the field of organic evolution. They are, however, by no means confined to evolution or to other biological questions, which include those of race. One has only to think of history, economics, and literature for other examples. How many high school students in Texas are told that some historians consider the defense of the Alamo a tactical blunder in the midst of a morally indefensible war? How many high school students anywhere in America are taught specifically that free enterprise has some grave drawbacks and socialism some great advantages? Or that *Lady Chatterley's Lover* is literature and why?

It is, however, pressures against the teaching of evolution that most concern me here, not only because that happens to be my own field, but also because I consider it the most important thing that needs to be taught at intermediate school levels. We are all familiar with the Scopes trial, if only from being reminded by the stage and movie success, *Inherit the Wind*. Many people seem to consider it as a quaint and amusing bit of ancient history that occurred in one isolated backwoods community. The fact is today that there are innumerable towns and whole cities that are just as opposed as Dayton, Tennessee, was to the teaching of evolution. And they are more successful in preventing it. I believe that most people misunderstand the serious issue in the Scopes case and, indeed, that it was misunderstood by most of the protagonists in the trial. There was a state law against teaching evolution in the public schools. Scopes broke that law, and the jury found him guilty. That was the only legal issue before that court, and it was correctly settled. That the verdict was upset on a technicality, that Scopes was not retried, and that the law is still on the books but has never been enforced and never had its constitutionality tested are all facts but beside the point.

What was actually argued in court, by prosecution and defense alike, was not the guilt or innocence of the defendant but the truth or falsity of evolution. There was, indeed, a social issue that transcended the rather trivial legal one. But certainly that really fundamental social issue was neither Scopes's guilt nor the truth of evolution. It was the competence of a legislature to enact and of a court to enforce the prohibition of teaching a theory that, whether true or not, was sustained by a large number of respectable scientists certainly competent in the pertinent field. By submitting the question of the truth of evolution to the court and jury, the defense equally with the prosecution compromised the whole situation and lost the one essential point. The point would have been the same if the law had made the teaching of evolution obligatory and Scopes had refused to teach it. Legislatures, judges, and juries cannot decide the correctness of a scientific theory or of the results of any scientific investigation. That can only be decided by further research in the self-correcting style of science. Furthermore, education will be stultified if properly qualified teachers are not free to teach what they believe to be true either from their own competence or on acceptable authority in the relevant field of research. That situation is also self-correcting. Any teacher who taught that the earth is flat would quite properly be discharged (not jailed!) for incompetence (not for breaking a law). Where there is evident unresolved conflict of authority, the teacher should of course explain that situation and may quite properly state his own position on either side.

Laws against teaching evolution are still nominally in effect over wide areas of the United States, but there has been no recent effort to enforce them. The prohibition is nevertheless now being applied far more effectively than by law and through agencies that are equally incompetent. They are incompetent in the usual sense of lacking the special knowledge necessary for rational judgment of the issue. They are also incompetent in a sense analogous to the technical concept of competence in law, that is, the competence of a court as having or lacking jurisdiction in a given case. The agencies now effectively prohibiting the teaching of evolution in many schools should have no jurisdiction over such a question. The competent agencies to decide on the subject matter of a science are the scientists and the science teachers.

Anti-intellectual control of science education by incompetent agencies is hardest to reach when it reflects the asininity of a local majority. "I won't have my child taught that stuff!" Such control may, however, be exercised by both vociferous minorities and individuals in key positions. There are also instances where community opinion is that evolution (or whatever the subject may be) is probably all right, but it is controversial, so we had better play safe and omit it from the curriculum.

At least one of the recent conferences of biological teachers had a formal discussion of this problem. (It has probably been discussed informally at all the institutes and conferences.) One suggestion was, of course, simply to sidestep, to omit anything about evolution one way or the other. Textbooks and the fact that teachers may have no voice in their selection make this easy. Some biology texts do omit evolution. Most of them relegate evolution to a single section, preferably in the back of the book, which need not be assigned. (There is little danger that students will read it anyway!) That also illustrates an indirect sort of censorship that can deny material to schools and students that would otherwise be receptive to it. If one community rejects the teaching of evolution and another does not demand it, some textbooks, at least, will aim for the least common denominator, and the chances are that neither community will get a book that treats the subject adequately if at all.

That solution, although probably the commonest one, is considered by many teachers to be dishonest. It cannot be intellectually honest to undertake to teach a subject but to omit its most important principle. It would, nowadays, be like teaching physics but leaving out atoms. (My high school physics teacher managed that, but that was long ago and far away.)

There was also mention of the possibility of teaching evolution but stopping short of man and making no mention of human evolutionary origins. This, too, can hardly be considered honest; and, in any event, it tends to cancel out the advantages of teaching evolution at all. It is neither necessary nor advisable to focus discussion of evolution primarily on man, but the main reason why teaching evolution is important lies in its implications for mankind. To omit even a glimpse of that connection would be not only to shortchange but also to mislead the student.

A third suggestion, apparently one that many teachers have already been acting on, is to teach about evolution but to leave out the dirty word. Call it "development" or "animal history" or the like. I gather that this has worked for some teachers, but it seems a transparent trick that is bound to be exposed sooner or later. It must cut down the coverage, too, for surely you cannot talk very much about "development" without letting the cat out of the bag and revealing that you mean evolution. I wonder, too, whether such teachers (and textbooks) are not being unnecessarily timid. Is it not possible that a system that will stand for teaching "development" will also stand for calling it by its right name?

Still another proposed, and actually used, solution is to present both sides of the case. Teach evolution under its own name as something that certain authorities believe. Also teach that certain other authorities do not believe it, and let the student decide for himself (or ignore the whole thing). This was hailed by some teachers at the institute as the most "honest" compromise on the problem, but I am afraid I cannot agree. It is less honest—because the student is less able to judge from data in his own hands—than teaching that some people say the earth is flat and some say it is round. It would be honest only if the teacher pointed out that the authorities who "believe" in evolution ("believe" is a misleading word here, too) are, almost to a man, those who have actually studied the subject in a scientific way and that those who do not believe in it are, almost to a man, obviously ignorant of the scientific evidence and swayed by wholly nonscientific considerations. That is not a compromise that would suit an antievolutionary school board. It might occasionally work in a controlling community that was open-minded about science but subject to some sniping from antievolutionary minorities.

The opposition to teaching evolution is, of course, almost always given a religious reason. That may usually be its real basis, but I think it is often a mask, perhaps unconscious, for underlying anti-intellectualism or antiscientism. Oddly enough, it is quite common to oppose teaching evolution on ostensibly religious grounds even in sects that do not in fact officially oppose or prohibit such teaching. Thus, many Catholic parochial schools are antievolutionary, but evolution is acceptable under Roman Catholic dogma and is taught in a straightforward way in many Catholic colleges. The whole situation is complicated by the fact that the dogma in some sects really is explicitly and violently antievolutionary and that some of these sects are highly evangelical, not only in religion but also in education. Some antievolutionary sectarian colleges specialize in science education, even in what there passes for biology.

If a sect does officially insist that its structure of belief demands that evolution be false, then no compromise is possible. An honest and competent biology teacher can only conclude that the sect's beliefs are wrong and that its religion is a false one. It is not the teacher's duty to point this out unnecessarily, but it is certainly his duty not to compromise the point. Fortunately, the great majority of religious people in America belong to sects that are more flexible on this point, even though the tendency of the average parishioner may be antievolutionary. Here a perfectly honest compromise, or rather a tolerant understanding, is possible. Evolution, per se, is not antireligious any more than the roundness of the earth is antireligious, although it was once held to be so. There are many religious and, in various sects, even highly orthodox evolutionists. There are also atheistic evolutionists, but so are there atheistic bankers, who nevertheless keep honest accounts. The lack of *necessity* for conflict between evolution and religion is something that can and, when the subject arises, should be pointed out by teachers. The most extreme and bigoted opponents cannot be placated, but there is plenty of common ground for reasonable people on this question.

There are, to be sure, many high schools where evolution is taught without opposition from students or community and even with their enthusiastic support. There are also textbooks that include evolution under its right name and as an established biological fact. Nevertheless, it is certainly true that innumerable students still leave high school without ever having heard of evolution, or having heard of it only in such a way as to leave them unimpressed or antagonistic. Since intermediate education is the proper level for encountering this subject and is for great numbers of people the only place where they are likely to learn anything valid about it, this means that an awareness of evolution is lacking or rejected in large segments of the adult population. Yet for over a century now, evolution has been known to be one of the great and central concepts of science and one fundamental for human orientation in the modern world. There is no other concept of comparable importance and scope that has been so slow in permeating education and in obtaining general popular acceptance.

That is what made H. J. Muller, on the centenary of the publication of Darwin's *The Origin of Species,* exclaim angrily that "One hundred years without Darwin are enough!" (I have pilfered that remark as the title of this chapter; I think Muller will forgive my theft in a good cause.)

Here, if not before, someone will want to ask, "Why make so much fuss about evolution? It is only one of a thousand things that might be taught in high school. Students can't learn them all. Naturally, you emphasize it because it is your specialty. A specialist in mathematics doubtless wants everyone to be taught calculus, but it isn't really necessary or even useful for all high school students to know calculus or evolution."

Part of the answer arises from personal reminiscence, again, for which further indulgence is asked. I do not think that evolution is supremely important because it is my specialty. On the contrary, it is my specialty because I think it is supremely important. I entered college with the intention of studying literature and becoming a writer, perhaps a poet. (Remember that my really enthusiastic high school teachers taught English.) I was required to take some laboratory science, and I elected geology, partly because of some previous interest in minerals and partly because Geology 1 was reputed to be a quick and easy way to work off the requirement. Actually, it turned out to be tough because a new professor (who did not last long in that college) demanded an amount of work that most of the students found excessive. But he was another enthusiast, and he imparted to me the thrill of learning things. Here I saw that it was possible to accumulate solid knowledge about the universe, new not only to me but to everybody, and to supply satisfactions that, for me, literary endeavor could not. I switched my major to geology. Slowly I came to feel that although minerals are fascinating, what is really important is life. That made paleontology, the living aspect of geology, my subject in graduate school. Starting then and increasing through the subsequent professional years, a sharpening sense of values showed me that if life is the most important thing about our world, the most important thing about life is its evolution. Thus, by consciously seeking what is most meaningful, I moved from poetry to mineralogy to paleontology to evolution. The transition would have been simpler if I had started with biology, or perhaps even with, say, chemistry; but I think the search would have wound up in the same place.

So I reached an answer to the suggestion that calculus and evolution are just two of many subjects and that no one can or should study every subject. Evolution is more important in an absolute sense, and it is important to everyone. Calculus, just as one example, is an excellent tool, indispensable in some quite specialized pursuits, quite irrelevant in others, and with no particular bearing on the human condition. It is evolution that can provide answers, so far as answers can be reached rationally and from objective evidence, to some of those big and universal questions I mentioned earlier. One has only to state some of the firm evolutionary generalizations and principles to establish their absolute importance and their necessary inclusion in a proper education for everyone.

1. Man is a recent and, up to now, in some real sense the highest product of a natural process that has been going on for billions of years.

2. Man owes all his characteristics to their gradual and very slow accumulation because they worked better, because they promoted most successful reproduction and continuance, through all the varying circumstances in which our ancestors existed.

3. The mechanisms and principles of that accumulation are now largely known and are probably entirely knowable in terms of the immanent physical laws of the universe. The source of those immanent laws themselves is quite unknown and probably unknowable to science; here religion may honorably enter the picture.

4. Knowledge of those mechanisms and principles makes it possible, within determinable limits, for man to influence for the better the further evolution not only of other organisms, but also of himself.

5. All living things are truly physically related in just the same ways as parents and children and brothers and sisters are related, although in greatly different degrees. In the enormously intricate and yet comprehensible pattern of life, man occupies a place unique to him but a place that is within that pattern and a part of it. Man belongs in and to nature just as much as any other kind of organism, and he is akin to all the others.

6. As a result of that kinship, man shares a great deal with all other organisms, most, of course, with his nearest relatives (in broad-

ening degree the apes, the primates, the mammals, the vertebrates), but much with living things as remote as trees or bacteria. We can learn much about ourselves in terms of processes in other species, much about them in terms of processes in ourselves.

7. Man's special capacities, his awareness, his perceptual functions, his reactability, his ability for symbolization and socialization, are all biological adaptations developed by evolution under the stress and guidance of natural selection. It is quite proper to speak of values in this process, and the values are inherent in the course and outcome of evolution. A working coordination between mental life and the outer world, a grasp on reality in the deepest sense, is one of the values required by and produced by our evolutionary history.

8. Our special abilities operate properly, which is to say in accordance with their natural functions in the evolution of our particular species, only if they are used rationally and responsibly. Rationality and responsibility are made possible and necessary by the evolutionary intensification of awareness and of flexibility of reactions.

9. Mankind *is* a kind, biologically a single species, united within itself and separate, as of now (although of course united through ancestral lines), from all other species. Like the members of any species, men vary. No two men are quite alike, and whole groups visibly differ, as the subspecies of widespread species always do. The resemblances among all men are vastly, incomparably greater than any differences. The more obvious differences arose, for the most part if not altogether, among early men as adaptations to particular situations and are biologically almost entirely irrelevant in modern civilization. There are no biologically superior or inferior races.

I could extend the list almost indefinitely, but I think that I have made my point, which is simply that evolution has fundamental human significance for everyone. Of course, I realize that such grand generalizations, presented just so, would be incomprehensible, incredible, or virtually meaningless for most high school students. Nevertheless, the implications are there, and some, at least, of them will eventually be glimpsed by anyone who acquires even a modest grasp of evolutionary facts and principles.

As to how to convey that modest grasp, I am no pedagogue, and I fall back on the disclaimer implied in my three forms of nonprofessional relationships with high school teaching. Of course, I do have some ideas on this score. (Teachers are like artists in that practically everyone feels competent to advise them without bothering to learn their profession!) Evolution underlies every aspect of biology and is one form of explanation for every biological fact, from protein synthesis to, say, zoogeography. As each topic is taken up, from the very first one—whatever that may be in the particular approach used—it can be shown to involve relationships best understood as results of evolution. Followed through, one topic after another, that builds up to a convincing demonstration of the fact of evolution. The first task is to show that evolution, as a general proposition, rests on good, solid evidence, and since *all* the facts of biology are evidence of evolution, that seems to me the way to approach the task. A routine listing of "proofs" of evolution as a short topic in itself can never carry such conviction.

With that general approach, specific information about processes and explanations of evolution, as distinct from (or, rather, additional to) the demonstration of the fact of evolution, will also emerge quite naturally. Most of the modern explanatory theory is inherent in the facts of genetics, ecology, and systematics if these topics are treated frankly in their relationships to the history of life. The broader implications, even though perhaps still on a more elementary level than those I previously gave as examples, will then begin to appear almost automatically.

I have no delusions that many teachers will soon and successfully follow the approach that I have suggested. Apart from all the elements of resistance against emphasizing evolution or teaching it at all, a great deal of ability and hard work would be required from the teachers—more than many of them can reasonably be expected to bring to the task. That task will, however, be eased by textbook and curriculum improvements that are currently under way and are among the too few encouraging signs. I must confess to certain reservations about the most elaborate and expensive of those efforts, the federally supported Biological Science Curriculum Study. None of its three alternative versions, in their present state, is oriented primarily on evolution and imbued with that principle throughout. All do, however, frankly introduce evolution under its right name, and one does emphasize the subject. Some of the less elaborate (and less diffuse) efforts of individual textbook writers are tending even more successfully in that direction. There are nevertheless still plenty of reactionary texts available for selection by antievolutionary school boards and teachers.

The millennium when all students will be given a good helping of modern knowledge of evolution is not at hand. I have stressed, possibly even overstressed, the difficulties and discouragements in the way. Still, perhaps an inch or two of progress will result if all those who see the opportunity and the necessity complain loudly enough about the present situation and put themselves productively and courageously to work.

Three Nineteenth-Century Approaches to Evolution

THE lives of Jean Lamarck (1744–1829), Charles Darwin (1809–1882), and Samuel Butler (1835–1902) spanned the turbulent nineteenth century, and they helped to make it turbulent. As different as three geniuses could well be, they nevertheless were alike in their passionate interest in organic evolution, and they were linked in curious ways. Their voices still are heard today, Lamarck's a feeble and distorted echo from the past, Darwin's a quiet but firm pronouncement by which we orient our lives, Butler's a shrill and petulant cry that has not lost the power both to entertain and to annoy.

The tenor of Lamarck's life was sad. The history of his ideas and reputation is ironic, at times so egregiously so as to become almost comic. In life, he was not without honor. He was, after all, early elected to the Académie des Sciences, and in all the uproar from Louis XVI through the revolution to Napoleon he held official posts that were among the most considerable then open to scientists. But he was honored for what were to him the wrong things, for accomplishments (mostly in the classification of plants and animals) that he considered, if not absolutely trivial, at least only incidental to his real purpose in life. He worked prodigiously in chemistry, geology, meteorology, botany, zoology, physiology, and psychology.

To him those studies were neither scattered nor disparate. They all were on the same subject: a world view, a cosmic philosophy that unified all phenomena. That aim was widely misunderstood in his day, as it still is now, and what was understood was rejected, as it must still be.

Lamarck lived to a great age, enfeebled, blind, and neglected. He founded no school and left no true followers. A few did some lip service to his memory, but none took up the task he had set and most repudiated him entirely, or worse, simply passed over his whole theoretical and philosophical labor in utter silence. Then after Darwin and *The Origin of Species,* and two generations after Lamarck descended to his limbo, some critics of Darwin's theory of natural selection developed an opposing theory that came to be called Lamarckism or Neo-Lamarckism. This is the ironic joke: that the theory to which Lamarck's name became and still remains attached and to which all his posthumous fame is due is fundamentally different from what he himself intended. It would have been bitterly repudiated by him, and he might well have preferred the neglect that was his lot while living. The few Neo-Lamarckians who read Lamarck inserted their own quite different conclusions between his lines. Most did not even read his work. As Guyénot, one of the very few historians who have indeed understood Lamarck, puts it, "Certain 'Lamarckians' have upheld concepts that are so far from that of Lamarck, so obviously contrary to what he explicitly stated, that they make one wonder whether they have ever read the *Philosophie Zoologique.*" The crowning irony is that Butler should have become Lamarck's most determined spokesman. "Who has such a friend does not need an enemy" might have been said of Butler's defense of Lamarck.

Just what, then, were Lamarck's views if they were not the origin of Neo-Lamarckism? The barrier to comprehension here is formidable. One must blot out the whole attitude of objective science toward the material world, an attitude that has now become almost second nature to western man, even (with some strongly defended exceptions) in the thought of nonscientists. For the fact is that outside of his routine but sometimes brilliant work in taxonomy

Lamarck was not a scientific pioneer, indeed not really a scientist at all, but one of the last of the subjective and deductive philosophers in direct line from ancient Greece.

Lamarck's philosophy descends in a noble lineage from Heraclitus. The fundamental aspect of nature is change. The world *is* process and activity, and its reality is less in fleeting material configurations than in the very fact of their fleetingness. Lamarck first applied this world view in chemistry, where he saw in fire, an element to him as to the ancient Greeks, the basic principle of activity and therefore of the whole universe. He was a resolute opponent of objective molecular chemistry, which was arising in his lifetime. He further applied that philosophical system to geology, where he denied the reality of minerals as species of fixed composition and saw in them only stages in a constant flux tending always toward disintegration and decomposition. As for life, it is the very essence of activity, with fire (or "caloric") as its physical basis or principle. Life is, moreover, the creative process in the universe. Through life the flux of matter leads upward continuously into higher states of activity, of "composition," or as we might put it—it is so difficult for us to leave the argument entirely in Lamarck's terms—into more complex compounds. As these products pass from living organisms into inorganic nature, the flux reverses and becomes a degradation toward the simpler and lower. The products of that retrograde process then pass again into living things, and so the cycle is ever renewed. Thus in a sense, and translated into somewhat modern terms, all chemical compounds are of organic origin, either in themselves or as stages in the decomposition of organic compounds.

As late as 1797 Lamarck, then fifty-three years old, mentioned in passing his acceptance of the orthodox eighteenth-century view that species of organisms are immutable. By 1800 he was teaching that fixed species do not occur in the organic any more than in the inorganic realm. He had a profoundly uniformitarian and monistic world picture, and he now saw that in *all* realms the rule must be flux and not configuration. In retrospect this has seemed to historians a radical change from belief in special creation to advocacy of evolution, a change, incidentally, considered remarkable in a man no longer young. I suspect, on the contrary, that to Lamarck there was nothing abrupt or revolutionary in this step. It was only a shift in attention that now brought out conclusions always inherent in a philosophy that he had embraced long since. By 1809 he had developed this into a coherent and specific philosophy of zoology in a book that bore that title, *Philosophie Zoologique.* He later restated and largely exemplified, but did not significantly change, that system in his *Histoire Naturelle des Animaux sans Vertèbres,* published between 1815 and 1822, at which latter date he was seventy-eight years old and slipping down to darkness and to death.

After a somewhat rambling introduction, Lamarck is first concerned in the *Philosophie Zoologique* with the proposition that species, genera, families, orders, and classes of organisms do not exist in nature. Darwin puzzled out the origin of species; Lamarck simply concluded that species do not exist. For Lamarck the whole mass of living things is a continuum and any subdivisions in the continuum must be artificial, for purposes of human convenience only. Nevertheless, there are dimensions in the continuum; its different areas lie at different distances from each other; their degrees of natural relationships differ. And there is always movement within the continuum. If the descendants of a given ancestry are followed through time, they change appreciably and, in the end, greatly. In short (and in a later terminology), they evolve.

The main movement of the mass of beings is upward in the sense of increasing complexity and ability to react. The "mass"— that is one of Lamarck's most characteristic but not quite explicitly defined concepts—moves up the ladder of nature. Its lesser parts, however, such as those that we artificially excise as species, suffer certain perturbations and do not invariably progress up the ladder in a single and straight line:

> I shall demonstrate that nature while giving, with the help of time, existence to all animals and all plants, has really formed in each of these kingdoms a true *ladder,* in relationship to the increasing complexity of organization of these living things, but that this *ladder,* which is to be recognized by comparison of the objects [material organisms] according to their natural relationships, has concretely specifiable

steps only in the principal mass of the general sequence, and not among the species or even among the genera [that is, in detail in smaller segments of the whole mass]. The reason for this peculiarity is that the extreme diversity of the circumstances in which the various races of plants and animals occur is not related to the increasing complexity of organization among them . . . and that that diversity causes the appearance of anomalies in forms and external characteristics or of aberrant species which the increasing complexity of organization could not in itself have brought about.

The whole of evolution then consists of two kinds of movements or fluxes. One is a continual (although, as Lamarck noted, not necessarily a constant) movement upward along the single route of the ladder of nature. It is, by the way, typical of Lamarck that although he considered this a progression, a perfecting, he frequently described it in reverse order, from top (end) to bottom (beginning), and then spoke of it as a degeneration. The other movements are local within the mass, carrying some of its smaller parts but never the mass as such in peculiar directions away from the straight line of progress. He often called these local movements "anomalies" and sometimes "digressions."

The grand mass movement, the basic flux of evolution, is continuously replenished at the bottom by spontaneous generation (Lamarck's own expression) of simplest organisms from inorganic matter, a process which is said to occur continuously every day. No sooner are they generated than those organisms join the mass movement and begin in turn to climb the ladder of life. Thus the simplest organisms of today are those that have been most recently generated and have not yet had time to progress further. The complex organisms of today are late states of organisms generated at earlier times. Those same higher states will eventually be reached by the organisms now being generated. The end of the mass ascent is man. Material compounded to this highest level does not evolve further but returns to the inorganic and is slowly degraded there until at the bottom it undergoes another spontaneous generation into life and starts the long, slow ascent toward man all over again. The whole mass of life, although always moving upward, never piles up at the top because the human population is self-regulatory and returns to the inorganic realm as much of the organic flux as rises to it from the organic mass below. Overpopulation is thus impossible. Malthus published his essay on population before Lamarck became an evolutionist, but Lamarck was no Malthusian—one of his striking differences from Darwin, although not among the most important.

A curious logical consequence of that view of evolution, duly noted and even emphasized by Lamarck, is that no group of animals, no taxonomically definable part of the ascending mass, can ever become extinct. Any local holocaust can only make a space that is immediately closed by the flow of the mass, like the obliteration of a hole in flowing water. (That simile is not Lamarck's.) Lamarck did admit the bare possibility that some digression, some anomalous group not in the mainstream, might become extinct through the action of man.

Another curious consequence, not so clearly noted by Lamarck although it is obviously inevitable under his theory, is that the whole of organic nature must now be in what in present terminology may be called a dynamic steady state. The main taxonomic groups of organisms (above the level of genera) must henceforth and forever remain exactly the same. Even though each single group, followed through its individual lines of descent, constantly moves onward, its space is not left vacant but is simultaneously occupied by others flowing up from below. In our previous figure of speech, the water flows but the river does not change. It may be questioned in what sense this view of a steady taxonomic state of nature can really be called evolutionary.

Lamarck was perfectly clear (although wrong) about the causes of the eddies and branch streams, his anomalies and digressions. This is the one part of his theory—and it is a comparatively minor part —that was remembered and often correctly represented by the Neo-Lamarckians. Peculiar circumstances of particular places where animals may find themselves face the animals with peculiar and particular needs. (The French word is *"besoins"*; mistranslation as "wants" or even "wishes" has been another falsification of Lamarck's thought.) To the extent permitted by their general organization (their place at that time on the ladder of nature), the animals respond by developing appropriate (adaptive) habits. Habitual use of an organ strengthens, develops, and enlarges it. Habitual disuse weakens, deteriorates, diminishes, and eventually obliterates an organ. The results of such use or disuse by individuals are passed on by heredity to their offspring.

According to Lamarck, and in keeping with his broader ontology, all that is significant in that process must come from within, from the organism itself as the vehicle or seat of the constructive flux of nature. Lamarck explicitly denied that direct effects of the environment were involved, or that they were heritable. The Neo-Lamarckians, all of whom accepted that factor and some of whom considered it paramount, here again were not being Lamarckian.

Lamarck was far from being so clear about the causes of the main, really important movement of evolution, the mass ascent of the ladder of nature. Passages can be found that seem, taken in isolation, to imply that this, too, is a response to ever increasing, ever more complex needs. Nevertheless Lamarck's most explicit remarks on the subject prohibit that inference. He repeatedly and definitely stated that there are two distinct causes of evolution:

> On one hand that [cause] special to the power of life in animals, a power which ceaselessly tends to complicate organizations, to form and multiply particular organs, finally, to increase and perfect faculties; on the other hand, the accidental and modifying causes (inherent in the very different circumstances in which animals occur), the products of which are the various anomalies in [digressions from] the results of the power of life.

In other words, the cause of the mass progression is the power, force, or capacity (the French word is *pouvoir*) of life; and environmental circumstances (acting through needs and hence use and disuse) produce only the minor divergences from that main line of progress. That says no more than that the main movement is an inherent characteristic of life, which certainly explains nothing. Lamarck wrote much else on the subject, but I, at least, am unable to abstract a clearer representation of his underlying concept. Perhaps one would have to be Lamarck in order to grasp it.

That vagueness and such expressions as "the power of life" have given the impression that Lamarck was a vitalist, but that is not true, at least in the sense of a dualistic distinction between vital and physical principles. Lamarck held that after the creation of nature by a divine Author the whole course of movement in nature has been entirely physical, while necessitated by the divinely created laws. Lamarck accepted and indeed strongly emphasized the inheritance of acquired characters as one of those laws, even though the role of the law in Lamarck's evolutionism was primarily in the rather trivial production of anomalies. That "law" is another concept as old as Greek philosophy, and surely far older. In Lamarck's day it was almost unquestioned and was generally considered self-evident. It still seemed virtually self-evident to Darwin, and the persistence of that error is one point—very nearly the only one—where there really is a historical, philosophical continuity that runs through both Lamarck and Darwin.

Darwin has been accused, most bitterly and even viciously by Samuel Butler, but also by other commentators even in our day, of cavalier treatment of Lamarck. According to all those criticisms Darwin did not soon enough or strongly enough acknowledge Lamarck's priority in the statement of evolutionary theory. Some, including Butler, went further and claimed that Darwin failed to make honest acknowledgment of a direct derivation of his ideas from those of Lamarck. In fact, Darwin's knowledge of Lamarck's theory seems always to have been extremely sketchy, and what early knowledge of it he did have evidently tended to put him off evolutionary ideas rather than to lead him to them—Lamarck had that effect on most biologists of his own day and the following generation. Stung by criticism, Darwin did include in later editions of *The Origin of Species* a summary of some of Lamarck's views. That summary is literally more than fair, because it abstracts just those points that, out of the original context, most nearly resemble parts of Darwin's thought, and it omits the more essential Lamarckian ideas that most of Darwin's contemporaries found ludicrous—and that, in all sympathy, truly are ludicrous when juxtaposed to Darwinian or to modern knowledge of evolution.

As a matter of historical fact the Darwinian, Neo-Darwinian, and modern synthetic theories of evolution (closely related among themselves) did not develop from the Lamarckian theory. With little qualification, the same can indeed be said of the Neo-Lamarckian theory. If, as far as now possible, one looks at the matter from a pre-Darwinian viewpoint, it seems that Lamarck's work could not have led directly either to general acceptance of the truth of evolution or to an understanding of real evolutionary processes. Lamarck's theory is at least as closely related to the philosophy of more than two thousand years earlier as it is to the science of only fifty years later. It is the contrast between the two theories and the two viewpoints, more than the resemblance, that is significant and interesting.

Darwin was no philosopher. He seems to have been quite unaware that his objective approach to the facts of life had implicit philosophical premises. He had no world view, no cosmic ontology, or rather he took the one he had for granted without introspective identification. His method, never formally learned but acquired by a sort of mental osmosis, was first to postulate some relationship among available facts, and then painstakingly to seek any additional facts that might check and more fully delineate the possible relationship. It was at this level that he learned that the basic relationships among all the groups of organisms are those of phylogenetic affinity or organic evolution. Simultaneously (for these two aspects were never clearly distinguished in Darwin's work) he sought some proximate explanation of those relationships, of the resemblances and differences and of the degrees of affinity involved in them. His explanation at this second level was natural selection, and that explanation, now enriched, is still seen as valid. Darwin knew there were other factors beyond this that up to this point had to be taken for granted. His only extensive effort concerning these factors was unfortunately oriented by a tradition that had reached him through Lamarck and many others, that of the inheritance of acquired characters. This one avenue of research in which Darwin was (albeit rather unconsciously) carrying on from Lamarck ended in a fiasco, a hypothesis on inheritance that Darwin's admirers would prefer to forget. (I will mention this again later.) Even here, although here without success, he was following his method of postulation checked and amplified by induction, and not the philosophical deductive method of Lamarck.

It is Darwin's method and not Lamarck's that is the method of modern science. Even had his theory been wrong—and we are now sure that it was right in the most essential respects—Darwin would deserve honor for the enormous step of bringing all of the phenomena of life into the field of objective scientific inquiry and inductive testability, as neither Lamarck nor anyone else had previously done.

Darwin's over-all view of the world of life is almost totally different from Lamarck's beyond the assumption by both that life evolves of itself and by the physical laws of the material universe. In Darwin's universe there is no complete lateral continuity, no ladder of nature, no ascending "mass," no contrast of anomaly and generality, no "power of life" with its undefined finality, no continuous cycling upward in the organic and downward in the inorganic, no steady state. Here life arose once, at a time and by means unknown but not unknowable. It then expanded and changed, branching into myriads of quite discrete lineages, many of which became extinct and were replaced not by their likes but by others better fit to occupy their places in the economy of nature. All this was in a world of constant change, progressive in the sense that each stage was built on and differed from all previous stages. The evolutionary movement was powered and controlled, not exclusively but in the main, by natural selection.

Darwinian natural selection was based on a few concepts all obviously true once they have been pointed out. After Darwin had pointed them out, honest biologists agreed they had been extremely stupid not to see them before. (Just one naturalist, Wallace, did see them without help from—and without in turn helping—Darwin.) All organisms vary, some being more and others less fit for survival. Much of that variation is heritable by their offspring. All organisms tend to produce more offspring than can possibly survive in the long run. On an average, more offspring will survive from those parents whose heritable variations make them more fit. Therefore, on an average and in the long run, characteristics that adapt various lineages of organisms to the different environments available to them will accumulate progressively within them. Q.E.D. The conclusion follows from the objective facts of nature as inexorably as the proof of a theorem in Euclid follows from his subjective axioms. We now see clearly what Darwin also sensed, but more vaguely, that the essential point is differential success in contributing offspring to the next reproducing generation and that individual survival is only one of numerous factors contributing to that result. This broadening of the concept has only enhanced the importance of natural selection.

Here we must anticipate and briefly answer two of the most clamorous criticisms of natural selection that were made by Butler and many others. First, they insisted that the real problem is the origin of the variations on which natural selection acts, and that nothing is explained unless the variations themselves are explained. Only now are we closely approaching an explanation of the origin of heritable variations. Darwin at first had no explanation and later provided a partial one that was wrong. Nevertheless the criticism as applied to Darwin was quite without force. Heritable variations do exist. That is an objective fact of nature known ages before Darwin and never seriously doubted by any rational person. A theory built upward on given facts may be entirely valid at its own level without any reference whatever to the origin of those facts. We would of course like to add ulterior explanation at each more elementary level, but to insist that this be antecedent to any higher level of theory is not only logically fallacious but also scientifically stultifying.

The second objection, which has become a cliché, is that evolution by natural selection is evolution by chance; and who can believe that the intricacy of living organisms and their delicate adjustment to the world they live in is the product of chance? There have indeed been biologists who believed in evolution by pure chance (some people can believe anything), but Darwin did not believe that and neither do modern evolutionists. The difficulty arises from the fact that variations do not seem to be predominantly oriented in the direction of the adaptations that organisms have in fact acquired.

Here there are a great many complications and subtleties, for the most part developed since Darwin. However, a complete and valid answer on the level of Darwin's own knowledge is clear enough. Natural selection is no random or chance process, but quite the contrary. It is directional and directive, and it inevitably must lead precisely in those adaptive directions that have in fact characterized the course of evolution. To be effective it does not require that all, or most, or many variations be in adaptive directions but only that *some* be, and it is a fact of observation that some are. One need seek no further for the antichance factor that must be involved in evolution.

Darwin never considered natural selection as all-powerful in the guidance of evolution. He always did consider it the most powerful of directive factors, at first to such an extent that when he wrote *The Origin of Species* he gave scant notice to any other possibilities. He was nevertheless sensitive to criticism that showered down not only from Butler but also from others of more orthodox scientific reputation. As we saw in Chapter 1, he weakened his stand in successive revisions of *The Origin,* and finally admitted four directive factors:

1. Natural selection, still, Darwin insisted most important.

2. Inherited effects of use and disuse (as in Lamarck's explanation of "anomalies"), important but secondary.

3. Direct action of external conditions, real but unimportant in relationship to adaptation.

4. What we now call mutations, also unimportant as a cause of adaptations. (Incidentally, the word "mutation" has had a curious history; it was already used, of course in a nonmodern sense, by Lamarck, and was also used by Darwin but not in our sense and not in this passage.)

The concessions here made by Darwin placed Butler in an ambivalent situation. On one hand, they enabled Butler to cite Darwin himself as authority for a part of Butler's otherwise quite antagonistic views, and they gave a specious excuse for Butler's insistence that some, at least, of Darwin's theory was cribbed from Lamarck. On the other hand, the very fact that Darwin had conceded an essential point that Butler claimed—even though not because Butler claimed it—somewhat blunted the edge of Butler's hatchet attack on Darwin.

Butler was known in his lifetime as *homo unius libri* and is again so known today, but the book is not the same. In his lifetime it was *Erewhon* (1872) and now it is *The Way of All Flesh*, published posthumously (1903). In fact he also wrote a number of other books, among them one proving that "Homer's" *Odyssey* was composed by a woman, and four on heredity and evolution: *Life and Habit* (1877); *Evolution Old and New* (1879); *Unconscious Memory* (1880); and *Luck or Cunning?* (1886). In producing that barrage of books, Butler was following his own advice to Erasmus Darwin (somewhat belated advice, as Charles Darwin's grandfather had died seventy-seven years before it was given): "'If his opponents, not venturing to dispute with him, passed over one book in silence, he should have followed it up with another, and another, and another. . . .'"

Samuel Butler encountered *The Origin of Species*—the first edition—when he was a young immigrant sheep farmer in New Zealand. Butler was at once converted to evolution, and "converted" is the right word. Nevertheless, he did not grasp what Darwin was actually getting at, and he still had not grasped it after some twenty-five years devoted in great part to this subject. He followed the words. He eventually saw, not with full clarity but adequately, what was meant by Darwinian natural selection and came to detest what he saw. What he never did comprehend was the attitude, the whole approach to the problem, the laborious amassing of facts, the self-discipline of submitting the most cherished hypotheses to the cold judgment of those facts, the admission of uncertainties, and the critical weighing of probabilities in the light of objective evidence. In short, what he did not grasp was science, its methods and its whole philosophical orientation. It aroused in him a feeling of repugnance, rather vague at first but eventually leading to a slashing anti-scientism.

Upon his conversion to evolution, Butler began to play with

the idea with the wry and imaginative curiosity that was one of his attractive traits. It was a game of "What if?"—the same sort of game he later played with the *Odyssey* and with a dozen other subjects. Is not Darwin's evolution mechanical? Then what if machines evolved, became in fact animate, and eventually outdid man in the struggle for existence? Or alternatively, what if machines were viewed as external organs, new limbs that man has evolved and that carry evolution beyond the limitations of flesh and blood? Both of those games were played in *Erewhon*, his first and during his lifetime his only successful book. However, the second "What if?" was then sketched only briefly, because, as Butler said, "There was more amusement in the other view."

Then came a third game of "What if?" and somewhere along the line—it is difficult to see just where—Butler forgot that it had all started in fun and began to take his game in earnest. What if limbs are corporeal machines, just as the machines are extracorporeal limbs? Then must not both kinds of machines have been designed?

Here Butler had a vast body of nineteenth-century thought to draw on. Its bible was William Paley's *Natural Theology*, published at the very beginning of the century (1802) and still tremendously popular in Butler's day. Butler studied it with at least as much attention as he devoted to *The Origin of Species*. Its theme is still a familiar one. Is not a watch, for example, obviously designed to serve a purpose? Then must not the incomparably more intricate and just as obviously purposeful human body, as well as the whole of nature, also have been designed? Paley's conclusion was of course that God was the designer. Butler accepted the whole argument, but his conversion to evolutionism forced him to deny the conclusion.

If not God, then who? Here Butler came up with the most stunning of all his "What ifs?" We know who designed the watch and all the other extracorporeal machines. What if the corporeal machines had the same designer, man himself? Taking off from that basis, Butler wrote his first serious, or perhaps semiserious book on evolution: *Life and Habit*. (The evolutionism in *Erewhon* was still consciously a joke.) Expressed as briefly as possible and not quite in Butler's sequence, the argument of *Life and Habit* is as follows:

We must know how to do a thing if in fact we do it. The fertilized egg in fact develops of itself into the adult organism. Therefore it must know how to do that. It seems improbable that this knowledge is conscious, but the things we know best of all are precisely those that we have learned so very well that we are no longer conscious of them. An example is the expert pianist who knows a composition so well that he is quite unconscious of all his individual movements that evoke the music from the piano. But the egg, as such, has never developed into an adult. To know how to do this so supremely well it must be unconsciously remembering long and repeated practice that occurred before the egg as such existed. The egg is not a new thing, not really a separate personality. It is just a detached part and a physical continuation of its parents, and hence of all its ancestors back to the first living blob of matter. They did the practicing and the learning, and the egg unconsciously remembers it all.

What of design? Clearly this is done by the organisms themselves. They meet new situations and feel new needs. They must then respond appropriately, learn something new, or perish. What they learn is remembered by their descendants. (Here, in the necessity of response to needs and inheritance of the response, is the tenuous connection between Butler's thought and Lamarck's.) And so, little by little, the more complex organisms, those with more fully stocked unconscious memories, are built up and finally produce man.

Butler has assured us that when he embarked on this thesis he thought of it as "an adjunct to Darwinism which no one would welcome more gladly than Mr. Darwin himself." If he really thought that (and when Butler speaks one never knows how much of the tongue is in the cheek), it emphasizes how completely Butler had misunderstood Darwin and the entire scientific effort of the times. Not only the thesis in itself but also the whole method of arriving at it and sustaining it are antithetical to Darwin and to objective biological science. Before publishing the book, Butler did suddenly perceive a part of the antithesis, apparently as a result of reading an attack on Darwin by the Roman Catholic biologist Mivart. Now Butler turned to Lamarck, whom he also misunderstood, and the completed book became flatly but still more or less courteously anti-Darwinian.

Butler certainly hoped that the scientists would take him seriously, but he was not quite sure whether to take himself seriously.

> I admit that when I began to write upon my subject I did not seriously believe in it. . . . What am I to think or say? That I tried to deceive others till I have fallen a victim to my own falsehood? Surely this is the most reasonable conclusion to arrive at. . . . Will the reader bid me wake with him to a world of chance and blindness? Or can I persuade him to dream with me of a more living faith than either he or I had as yet conceived possible? As I have said, reason points remorselessly to an awakening, but faith and hope still beckon to the dream.

Butler prefaced those final remarks in *Life and Habit* by saying, "I am in very serious earnest, perhaps too much so," but it is he who goes on to question his own motives, to invoke faith and hope against science, and to call his theme an irrational dream. That is characteristic of Butler's perverse brilliance. Not only here but repeatedly in his numerous works he says in effect, "Frankly, I have been talking nonsense, and now if you call it nonsense I am still one up on you. Just the same, I expect you to believe what I said before and not to believe me when I call it nonsense."

Butler wanted and may even at times have expected his theory to supplant Darwin's and to transfer to him the great scientific fame of which he became openly and spitefully jealous. But at the same time, and this was not unrelated to the jealousy, he despised science and scientists. One of his many responses to that ambivalence was clever but cruel. He maintained that there are two kinds of "scientists." One kind simply knows instinctively how to live, without learning it; if they are illiterate so much the better. These "scientists" have "good health, good looks, good temper, common sense, and energy . . ." and are thoroughly delightful people. The others, the ones that are currently called scientists, ". . . are ugly, rude, and disagreeable people. . . ." It is the duty of the handsome instinctive "scientists," those who simply "know what's what," to adjudicate on the findings of the ugly, grubbing kind of scientists. So Butler places himself in the first class, although with some feigned misgivings (after all, he does know how to write and that is wrong for this kind of "scientist"), and he adjudicates on Darwin, plainly one of the rude and disagreeable kind. This is a rather mild example of the many times that Butler wooed scientists with brutal abuse.

Butler's second book on the subject, *Evolution Old and New,* purports to be a history of evolutionary theories. One of its theses is that study of evolution went seriously astray when it fell into Mr. Darwin's hands. (*Mr.* Darwin is Charles.) The good and true line runs through Buffon, Dr. Darwin, and Lamarck, to Butler. (*Dr.* Darwin is Erasmus, Charles's grandfather.) Nevertheless Mr. Darwin derived all his ideas from Dr. Darwin and Lamarck, while Butler's ideas were only in the smallest part adumbrated by Buffon and Dr. Darwin. The contradiction is typical of the self-consciously tricky book. We need follow it no further than to mention two amusing examples of Butler's clever perversity.

Buffon flatly said that animals have no knowledge of the past, no idea of time, hence no memory. That is the exact opposite of what Butler wants Buffon to have believed, so Butler concludes that Buffon must have been "laughing quietly in his sleeve." We can hardly complain that Butler is misrepresenting Buffon, because he says quite frankly that he is, but he does so with the evident hope that we will not believe it.

Butler does admit that Lamarck really did mean things that Butler cannot support, but at least Lamarck had some good ideas and did better than Mr. Darwin. "Lamarck would not have hesitated to admit, that, if animals are modified in a direction which is favourable to them, they will have a better chance of surviving and transmitting their favourable modifications. . . ." In other words, Lamarck would have anticipated Darwin on natural selection if that idea had occurred to Lamarck! The general conclusion is that Darwin is not only wrong but also a mere plagiarist.

In the next book, *Unconscious Memory,* Mr. Darwin becomes

not only a plagiarist but a liar. The latter accusation grew out of a misunderstanding too silly and too complex to explain here. Those interested in this tempest in a Victorian teapot will find all the pertinent documentation in an appendix to Lady Barlow's edition of Darwin's autobiography. Suffice it to say that the incident was a symptom and not a cause of Butler's now implacable hatred of Darwin. Darwin was for once stung to the quick and wanted to explain his position publicly, but was dissuaded from doing so by friends and relatives. His silence of course goaded Butler to greater fury than any reply could have done.

In addition to discussion of that incident, *Unconscious Memory*, a hodgepodge, follows the history of Butler's views and his writing of the first two books, and it then considers the related theories of two German biologists.

The fourth and last of Butler's books on evolution, *Luck or Cunning?*, is a rehash of much already said and a continuation, now thoroughly ill-tempered, of the personal attack on Darwin, dead five years since and unable to reply if he had wished.

After reading all of Butler's work on this subject, one is tempted to dismiss him as a mere dilettante insanely jealous of Darwin's reputation and acting like a cur snapping at heels. Darwin himself, in a passage not meant for publication, summed up with a quotation from Goethe by way of Huxley: "Every whale has its louse." That is both the most conceited and the most unkind remark that has reached us from an extraordinarily humble and kindly man. It might even have pleased Butler, for it shows that his darts had indeed made wounds.

It would, however, be both unfair and inaccurate to leave matters at that. There are here two broader issues that transcend those of personal opinion and antipathy. First, Butler had identified a real weakness in Darwin's theory and a basic problem that must be solved. Butler himself, however, approached the problem in the wrong way and left it farther from solution than ever. The second broad issue arises from the fact that Butler's failure illustrates and in a sense was caused by a damaging cultural cleavage that has become even more acute since his day.

It is a fact that the past experience of a race must to some extent and in some way be communicated to each individual born into that race. The experience consisted of gradual adaptation to particular circumstances and frequently to changing circumstances. Darwin solved, not in all detail but in its basic essentials, the problem of how the adaptation is brought about. He did not solve the problem of how it is communicated to descendant organisms. Butler was right to insist that that problem, too, must be solved before we can fully explain evolution, but Darwin was quite as aware of that necessity as Butler.

Darwin also attempted a solution. His suggestion, which is called pangenesis, was that particles from every region of the body are carried to the germ cells and that they transmit what is necessary of the ancestral experience. That idea is just as wrong as Butler's unconscious memory and, incidentally, it was the one idea of Darwin's (aside from the mere fact of evolution) that Butler found congenial.

As Darwin and Butler were equally wrong in regard to the particular problem that was central to Butler, it might be supposed that there is nothing to choose between them. That is decidedly not true. In the first place, Darwin had solved the problem that needed prior solution, that of adaptation. By rejecting Darwin's successful solution of that point, Butler closed the door on any possibility of further successes. In the second place, Darwin's method was right. He set up a hypothesis and painstakingly sought evidence bearing on it. He did not find adequate evidence; the time was not ripe. Pangenesis remained for Darwin not a theory, for the error of which he could be reproached, but an unverified hypothesis. "An unverified hypothesis," Darwin wrote in precisely this connection, "is of little or no value. But if any one should hereafter be led to make observations by which some such hypothesis could be established, I shall have done good service. . . ."

In fact by following Darwin's scientific approach, although not his original hypothesis, which the new observations disproved, the problem has now been largely solved. Such experience of the race as is passed on is, for the most part, in the form of coded information carried by an identified chemical (called DNA for short) associated with the chromosomes in the nuclei of the germ cells. Natural selection, in its modern guise, adequately explains how the information becomes coded simultaneously with, so to speak, its acquisition.

Where Darwin failed to reach a solution by the right method, Butler reached a wrong solution by the wrong method. Butler's own statement, more than amply supported by his performance, was that his approach, methods and intentions were literary, artistic, non-scientific, and even antiscientific. Basically his whole argument is a play on words. One may say, if one wishes, that sodium "knows" how to combine with chlorine to make salt, that an embryo "knows" how to develop into an adult and that a pianist "knows" how to play a composition. But the nature and source of the "knowledge" are so completely different in the three cases that use of the same word, "knows," becomes only a metaphor or a pun. Butler's uses of the words "memory," "learning," and others suffer the same fate. Butler mistook his word game for truth.

(Here I would like to insert a parenthetical cautionary query. Are we always quite sure that we are not playing one of Butler's word games when we use such terms as "information," "coding," and so forth in discussing these same matters today?)

Now, there is nothing whatever wrong with being literary, artistic, or nonscientific—we might add for Lamarck's benefit that there is also nothing wrong with being a deductive philosopher. Butler's works that are genuinely and appropriately artistic can still give great pleasure while his pseudoscientific books only annoy. The error lies in applying artistic methods and judgments to inappropriate fields. When Butler says in *Life and Habit*, "I know nothing of science, and it is as well that there should be no mistake on this head," we feel that he does not mean it but that it is true just the same. Yet the book is intended to correct those who do know something of science, and to do so in their own field. Butler claims for the artist the right to "adjudicate" on science.

The fact is that Butler hated and feared science and scientists, even while he envied them. He did not reject natural selection because he found any compelling opposing evidence. On that subject he adduced only irrelevant, although sometimes quite diverting, literary ridicule. His rejection was entirely emotional. He fought the idea of natural selection because he did not like it. Fearful and wonderful as he was—and as we all are in very truth—he did not want that marvelous being to have originated in so prosaic a way.

That sort of thinking is still with us. Shaw, a great admirer of Butler's, issued thunderous pronunciamentos against natural selection, with (if possible) even less consideration of any pertinent evidence. A recent literary biographer of Darwin concludes that by great good fortune natural selection is no longer a tenable theory. In fact its scientific status is now quite unassailable, as the lady would have known if she had been one tenth as knowledgeable in science as in literature. Perhaps there is a world where wishing makes things so, but it is not the world of science, nor indeed any real world we can live in.

C. P. Snow has recently given us a catchword for the basis of those antagonisms: it is the contrast of "the two cultures." The contrast is older than Darwin, or Lamarck, or Newton. It was, however, particularly exacerbated when science finally brought *all* the phenomena of life into its sphere, and that was Darwin's doing. Darwin himself was content in his own culture, largely indifferent to the other except for flinching occasionally when someone threw a rock from the other side. Indifference and incomprehension are still more characteristic in the other culture, but once in a while someone there, like Butler, takes a look, loses his indifference, and begins throwing rocks. I greatly hope that contacts between the two cultures will increase, but I doubt whether throwing rocks, from either side, is the right kind of contact.

CHAPTER FOUR

A Modern Approach to Evolution

IT is a pity that Butler's ideas were so infused and compounded with nonscientific nonsense and antiscientific polemics. They only confused issues and had no useful sequel of clarification. Yet if certain of them could have been put in a more sensible context, seen in a different light, they would have been prophetic and might even have led to some scientific insights achieved later and without indebtedness to Butler. It is a mistaken metaphor or a merely senseless, however amusing, pun to say that an egg "knows" how to develop into a hen because it "remembers" the past experience of the race. Nevertheless, as was mentioned in passing in the last chapter, it is true that an egg contains some sort of program or information, in an unconscious and impersonal sense, capable of converting egg to hen. It is also true that this something in the egg does somehow embody ancestral experiences. Clear recognition of those facts came to evolutionists long after Darwin and Butler. Finding out just how the experiences affect the egg and the way in which they are, so to speak, recorded there have been major problems of twentieth-century biology. Solution of those problems is a triumphant theme of recent research on evolution.

Butler also said that a hen is an egg's way of producing another egg. That is one of those remarks that seem very clever, even profound, at first sight but that turn out not to mean much if you think them over carefully. It is like the old and fundamentally senseless question as to which came first, the hen or the egg—senseless as a question because either answer is wrong. There are, nevertheless, two different points of view involved here, and while neither one is right if taken by itself and considered complete, it is also a fact that neither one is wrong if they are taken in conjunction. There is a hen-evolution and there is an egg-evolution. They are not at all separable things and neither has priority because they are different aspects of a single process.

Naturalists have always been interested primarily in hens rather than in eggs. If forced to take such a stand, they would counter Butler by saying, with equal truth and equal inadequacy, that an egg is a hen's way of producing another hen. Darwin, a quintessential naturalist, was basically concerned with hen-evolution. Of course he knew and illustrated with many examples that what the hen—that is, any adult animal—is and does is often determined and always circumscribed by the egg—that is, heredity. Yet the main thing, the focus of interest, was the hen. Although he did not clearly see or put it in this way, Darwin's theory of natural selection was the first and most important step toward understanding how hens' experiences affect eggs. The Neo-Lamarckians were also hen-evolutionists, and in this respect their point of view was the same as that of the Neo-Darwinians. They merely had a different, simpler, and, as it turned out, completely false idea of how hen-evolution got into the egg.

Early twentieth-century geneticists were more nearly of Butler's mind in the one respect that they were more interested in eggs than in hens. They concentrated on the half-truth that what the hen is and does is a result of egg-evolution. They believed that evolutionary changes in hens were wholly consequences of the fusion of the sex cells, gametes, to form a fertilized egg, and of events in cells that are precursors of the gametes. Those processes seemed to have no particular relationship to earlier events among hens, and indeed the hen (the man, the tree) seemed largely irrelevant except as a means of producing eggs.

It has become obvious that neither hen-evolution nor egg-evolution is the whole story. They must be considered together, and in fact the most crucial part of evolution is not in one or the other but in their mutual interaction. One modern approach to evolution is, then, to view it as a synthesis of the naturalists' hen-evolution and the geneticists' egg-evolution. This synthetic theory also synthesizes a great deal more. Every aspect of modern biology, from molecules to communities and beyond, enters into it. I am personally completely convinced that this theory is essentially correct over-all and as far as it goes, although I am well aware that it is not complete in detail and is also subject to correction in detail. Most students of evolution now agree, but it is inevitable and indeed desirable that there is a minority of dissenters. Some of the minority views are discussed elsewhere in this book. The fact of dissent is here mentioned in all fairness, but for the rest of this chapter I shall present only the views that I believe to be correct.

The early geneticists and those still belonging to what may now be called the classical school used and still use the hen mainly as a means of finding out what is in the egg. Some of them literally study hens for this purpose, but (as is well known) fruit flies (species of *Drosophila*) are more practical and have been more widely used. Many other organisms have also been studied from this point of view. In recent years microorganisms are increasingly used: especially bacteria but also fungi, algae, and protozoans; also viruses, which in my opinion do not really qualify as organisms but do have part of the reproductive apparatus stripped down to rather less than its bare essentials.

Earlier in this century those studies demonstrated that heredity is conveyed, for the most part, by some kind of unit to which the name "gene" was given, and that the genes are distributed in single file along the threadlike or rodlike chromosomes in the nucleus of each cell. It was also learned that in the process of maturation of the sex cells (gametes) segments of chromosomes with their genes could be rearranged or switched from one chromosome to another—recombined. Most animals and plants have two sets of chromosomes in their body cells and the single set in each gamete is derived at random from those two sets. A male gamete and a female gamete combine to give a double set in the fertilized egg that develops into an adult organism, which thus has another kind of recombination of chromosomes, hence genes, derived from its parents. Many, probably most, and perhaps all genes occur in different forms, called alleles, recognizable by their effects on the developing organism. Changes from one allele to another, called gene mutations, occur spontaneously at low but fairly constant rates typical for each gene. The rates can be artificially speeded up by various forms of radiation and chemicals.

The earliest studies were made, and many still are, on single genes where alleles produce some one readily visible effect. One allele produces a fly with red eyes and another a fly with white eyes, one a fly with normal wings and another a fly with stumpy wings, and so on. Later it was found that each gene affects many characters and each character is affected by many genes. Moreover, genes interact, they modify each other's actions, and although each is some sort of unit in itself, they act in common as parts of a unit of a higher order, an integrated system involving all the genes of all the chromosomes. What the system does is to control development from fertilized egg through embryo and young to adult and the functioning of the cells of the organism throughout.

In some respects genetic control is very rigid; for example, human blood types are determined once and for all when the egg is fertilized. For other characters, a majority of characters in most organisms, there is a certain amount of play, so to speak. The genetic system determines a larger or smaller range within which the organism will fall, and just where it is in that range is determined by environmental influences. For instance, the genetic system determines how tall or how short a man *may* be, but within that range the exact height that he *is* depends on nourishment, illness, and other nongenetic factors. The actual characteristics of a real organism are thus the outcome of interaction between its inherited genetic system and its experiences during life from the fertilized egg (technically a zygote) onward. The old argument about heredity versus environment is usually based on false alternatives, because as a rule both are involved.

Chromosomes are easily visible with ordinary microscopes, and in favorable cases enough of their structure can be seen to observe when segments are duplicated, reversed, or recombined in various ways. Genes, however, are not visible as such even with an electron microscope. Until recently they were theoretical constructs. They *must* exist because there is no other explanation for what goes on in developing organisms. By ingenious experiments it was even possible to determine very closely where many particular genes are located in certain chromosomes of given species. Their effects on the organism were also determined in great detail and in many cases— effects, to be sure, often recognizable only as indirect results of the interaction of numerous genes, and yet sometimes quite direct and specific. In any case the effect was generally quite clearly interpretable in terms of gene action. Nevertheless it still was not known just what genes were and just how they produced their effects.

What genes are has been discovered in the last few years, and how they produce their effects is now known in very small part and promises to be better known within the next few years. It has been learned that the material basis of heredity is a family of large molecules called nucleic acids. In most cases the primary active carrier of heredity is a deoxyribonucleic acid, mercifully abbreviated to DNA, located in a chromosome. In figurative terms (and let us not forget that this is figurative) the hereditary message, the information for development and cell function, is coded in DNA molecules. For our purposes we need not go into the exact nature of the coding, which is now almost but (on the day of writing) not quite fully known. It involves the sequences of four subunits of the molecule, and those subunits, like letters or other symbols, can be arranged to convey definite and different "messages."

DNA has three extremely remarkable properties essential for the process of heredity. First, of course, is that coding property, the fact that the subunits can be arranged in an almost endless variety of sequences, each highly specific and unique. Second is the fact that each DNA molecule, in the environment of a living cell, can direct the formation of duplicates of itself, new DNA molecules with the same specific sequence of subunits. That is a necessity not only for passing on the specifications of heredity from parents to offspring but also for proliferation of cells, each with its own chromosomes, from a single cell to the trillions that make up such higher organisms as men or oak trees. The third property of DNA is that it can pass on its sequential pattern or "message" to another kind of nucleic acid, RNA, which in turn and in a complex way can bring about the synthesis of a unique protein, indirectly specified by the original unique pattern.

What, then, is a gene? It is simply a segment of a DNA molecule that specifies a protein (or a significant part of one) indirectly through that complex process involving also RNA. A gene mutation is simply any kind of change in the subunit sequence of the DNA segment that is a gene. Since the gene segment is long and has many subunits, many such changes of different kinds can occur in it—up to several hundreds. That explains why a gene can have numerous alleles, although presumably many of the possible changes just scramble the "message," so that it does not specify a proper protein at all, and so knock the gene out as a functional unit.

These facts also show how a gene acts at the most basic level. It indirectly causes the formation of a specific protein which can act as an enzyme, bringing about specific further chemical reactions in the cell. Together with the initial structure and the chemistry of the cell, those enzymatic reactions determine how the cell will function and how, in a multicellular organism, a fertilized egg (a cell that is a zygote) will grow into an adult. That glib statement glosses over some colossal ignorance, for in fact the chemists are close to connecting gene and enzyme but as yet have not made even a good beginning on connecting enzyme and developed organism. In this approach, we know a great deal more about the egg than we do about the hen.

Later in this book (Chapter 6), I have something to say about different approaches to biological problems in general, with stress on the fact that all approaches are necessary and no one is sufficient.

Here I have just been considering, in of course an extremely summary and elementary way, the greatest triumph of the biochemical approach so far. It is in no way undervaluing these exciting, truly epoch-making accomplishments to point out, first, that this approach would not have led to those accomplishments if organismal biology had not pointed the way, and second, that the DNA studies have not yet contributed much to the fundamental study of evolution. The nucleic acids were known to chemists for many years before they were more than just curious compounds of unknown significance. It was studies of heredity in whole organisms that revealed their significance and told the chemists what to look for. And long before it was known that genes are segments of DNA it was known that genes exist and, to great extent, what they do. Knowledge of DNA has as yet cast almost no further light on what genes do *in terms of the organism*. Surely such light will eventually be achieved, and that will be another great step forward, but in the meantime we must keep a sense of perspective and not imagine that the chemical approach has already solved any basic problems of evolution or is capable of ever doing so alone.

The fact is, of course, that what I have said about heredity so far does not involve evolution at all. It has just been an abstract of parts of the story of how eggs acquire a developmental mechanism that causes them to become hens very much, but not exactly, like their parents. The facts of genetic coding and reproduction tell us why hens are more or less like their parents. The facts of mutation and recombination tell us why hens are never *exactly* like their parents. But none of those genetic facts tell us how the parents became hens to begin with, or how the offspring may eventually become something else—and of course that is what we are after. The genetic principles so far discussed are a necessary basis for understanding evolution, but we are only now coming to evolution itself.

The next step in our approach (which does not follow the historical sequence in which the facts were discovered) is to ask what it is that actually evolves. Early genetical studies seemed to imply that it is individuals. Attention was focused on single mutations that produce strongly marked effects on the individuals in which they occur. Those individuals are distinctly unlike their parents, so it could be said that in them an evolutionary change has occurred. The mutationist school, mentioned in the first chapter, concluded that this is all that is really essential in evolution. That idea still has a strong hold on popular imagination. The sudden appearance of the next species of man, a *Homo superior,* as a mutant among us poor *Homo sapiens* still often recurs in various guises in science fiction and even occasionally in supposedly nonfictional popular science writing. Although some science-fiction writers claim prophetic powers, they are in this respect (and in a number of others, too) a generation or two behind the times.

In retrospect it seems obvious that the mutationist view of evolution simply cannot be correct in a general and adequate way. The mere fact that an individual differs from its parents does not mean that evolution has occurred in a meaningful sense or to a significant extent. *All* individuals differ from their parents, some more and some less. Even without mutation, the various kinds of genetic recombination see to that. In any species, the possible number of recombinations of genes (with their various alleles) vastly exceeds the total number of individuals that have lived, are living, or ever will live. With the unimportant exception of identical twins, the chances that any two individuals will be genetically identical are virtually nil. The extensive variation among individuals of all species is necessary if evolution is to occur, but it does not in itself constitute evolutionary change. As a rule, recurrent mutations in any species simply add to that variation or maintain its level without, in themselves, leading to the formation of new species.

The second serious inadequacy of mutationism is that it was based on exceptional mutations, relatively rare in themselves and still more rarely likely to produce lasting evolutionary effects. Most mutations do not produce large, readily visible effects comparable to the distinctions between different species. The effects are usually quite small and are comparable, rather, to the usual variations within a single species. It is, indeed, probable that many mutations produce effects not noticeable at all by the usual methods of genetic analysis, although obviously that can only be checked either by the development of more refined genetic methods or by the as yet unattained chemical determination of actual mutations in DNA. Moreover, the greater the effects of a mutation, the less likely it is to have evolutionary significance or, in a sense, value. We have noted that the whole genetical system of an individual is a delicately adjusted, interacting, integrated unit. Introduction of a radically different element into that system will almost always upset its delicate integration and produce an ill-adapted or monstrous individual or quite likely prevent any development at all of the individual.

The third and most serious deficiency of mutationism as the sole or leading factor in evolution is related to that last fact. The most striking thing about organisms is that they are all intricately adapted to their particular ways of life in their particular environments. There is much more to say about that in this and in later chapters. At this point we need only note that mutations are not oriented toward such adaptation. In that respect they are random, and if they, alone, produced evolutionary change, evolution would be the result of mere chance. But that is absolutely incredible. It just is not possible that such thoroughgoing, almost inexpressibly complex adjustment between organisms and environments was produced by chance. That relationship must somehow have affected the genetic system as the organisms and their adaptations evolved. What happens to the hens must somehow and in the long run affect what is in the eggs. In terms of cybernetics and information theory, there must be some form of feedback from the organism-environment interaction to the genetic system. Mutationism, alone, provides for no feedback, which must therefore be sought elsewhere.

Neo-Lamarckism postulated a simple and direct feedback in individuals, but that does not really occur. Evolution does not happen in individuals, but in *populations*. That should have been obvious from the start, because evolution is an ongoing thing, through the generations and the ages, through some two billion years now or perhaps even more. Its continuity can only be through the populations that endure while individuals pass, and those populations are the medium in which evolutionary processes, including the feedback we now seek, must occur.

The population is a continuing, integrated unit and a bounded system, as is the individual but in a very different way. The continuity of the individual in time is brief and limited to that of the body as a discrete object; its integration is a physical union and direct interaction of parts; and its boundary is also physical and obvious, a continuous membrane or epidermis. The continuity of populations is potentially limited only by that of their environments, and it persists not in one discrete object but through the lines of heredity passed from such objects, parental individuals, to others, individual offspring. The integration of populations is not within but among discrete objects. It exists at several levels, such as that of behavioral interaction, but most fundamentally in a mesh of heredity as genetic factors from different parents come together in offspring and separate again in later generations. The population boundaries, not always absolute or clear, are set by the limits to which the genetic interchanges extend. The natural unit of evolution, then, is a continuing population, a lineage, the members of which interchange genes through the generations. The usual form of interchange is by interbreeding among males and females, but other forms of interchange also occur in organisms, such as bacteria, that do not have true sexes. (There are organisms that do not interchange genes at all, but they are exceptional.)

Such an interbreeding (or otherwise gene-interchanging), continuing population or lineage is a species, as species are now best defined in genetical and evolutionary terms. At any one time, such a species has a characteristic genetic composition, its genetic pool, made up of the totality of chromosomes (usually) and genes (in chromosomes or, very exceptionally, in other forms) of all members of the population. The makeup of the genetic pool is conventionally expressed by percentages, relative frequencies, of different forms of particular chromosomes or alternates (alleles) of particular genes present. The pool is not constant through any considerable period of time. It fluctuates from generation to generation, and also frequently has long-range trends of change. Those changes in genetic pools are the basic feature of evolution. In a sense you can say that they *are* evolution, although that becomes a rather meaningless abstraction if it is not related to the developing and developed organisms whose nature is largely determined and wholly limited by their genes.

The fundamental factors of evolution, then, are those that produce changes in the relative frequencies of chromosomes and genes in the genetic pool. First a surprising fact should be noted: the constant recombination of genes within chromosomes and of chromosomes drawn at random from the (usually) double parental sets does *not*, in itself, change the genetic pool. It maintains variation among the individuals of the population and ensures that any two of them are extremely unlikely to have the same genetic constitution, but the relative frequencies in the whole population are not affected. That is a basic principle of population genetics, known as the Hardy-Weinberg law after two geneticists who worked it out independently as a necessary statistical result of the known processes of gamete formation and sexual reproduction. Its interest is largely theoretical because there are always other factors at work that do tend to change the pool, but it shows that we must look elsewhere for the causes of evolutionary change and can eliminate recombination in this respect.

The main factors of genetic change were mentioned in Chapter 1 and must now be considered somewhat more fully. They are: (1) mutation, (2) sampling errors, (3) genetic migration, and (4) natural selection.

We have already seen that if mutation alone governed evolution, the usual outcome would be disadvantageous. Inadaptive, monstrous, or unviable individuals are often (but by no means always) produced by mutation, and if that were the usual outcome in a population, extinction would be the eventual result. A single progressive step can occasionally occur by chance, but that is clearly uncommon and probably never leads to sustained progressive evolution. The randomness of mutation with respect to progressive evolution or adaptation obviously must be counteracted somehow. One counteracting influence, although it is now believed to be a relatively minor one, is the fact that a mutant allele can back-mutate. That is, if an allele A_1 of a gene mutates to A_2, A_2 may eventually mutate back to A_1. Rates of back mutation are always much lower than those of the original mutation, and it is now known that they are even lower than geneticists used to think. A_2 is very unlikely to mutate to A_1, but may mutate to still another allele, say A_3, that has much the same effects as A_1. In any event, as mutation from A_1 to A_2 increases the frequency of A_2, that will eventually be balanced by mutation of A_2 back to A_1, or to A_3 or some still other allele, and the continuing frequency of A_2 will then tend to remain constant.

The major influence counteracting the random effects of mutation is, however, natural selection, the fourth of our pool-changing factors. As will be discussed soon, selection increases the frequencies of the relatively few adaptive mutations and decreases the frequencies of the more numerous inadaptive mutations. It should be emphasized, as also more fully discussed elsewhere, that mutations are not necessarily per se good or bad: whether they are adaptive and selected for or inadaptive and selected against often depends on the circumstances in which they occur and their combinations with other genes rather than simply on the character of the mutation as an isolated event.

The positive role of mutations in evolution, and it is a highly important one, is that they introduce *new* genetic factors into the pool, factors not simply derived by offspring from parents. They thus provide raw materials on which selection works.

"Sampling error" is a statistical term applied to chance fluctuations in the genetic pool from generation to generation. If the parental genes passed on to the offspring were in the exact proportions of the parents' genetic pool, the offspring's genes would constitute a fair or unbiased sample drawn from that pool. There would then be no evolutionary change. However, it is extremely unlikely that the chromosome sets drawn at random from male and female parents will have exactly the same relative frequencies as in the genetic pool. Quite by chance, the frequencies in the offspring differ to some extent from those in the parental pool, and that is what is meant by sampling error.

Sampling error always affects populations to some extent, and there are always some random fluctuations in genetic pools from generation to generation. In large populations the fluctuations tend to be relatively slight and to cancel out in the course of time. In small populations the fluctuations are more important and may become irreversible. That is, when the proportions of alleles A_1 and A_2 are fluctuating, one may eventually occur in all individuals and the other in none. Return to earlier relative frequencies is then obviously impossible.

Like mutation, sampling error is random with respect to adaptation. In itself it can only very rarely lead to a progressive single step in evolution and probably never to a continued trend. Like mutation, also, its random effects are usually counteracted by natural selection. There is no doubt that sampling error does occur, but there has been much controversy as to its over-all importance in evolution. The present consensus is that it is usually overbalanced by selection, that is, it rarely leads to elimination of a genetic factor favored by selection or fixation of one opposed by selection.

Genetic migration is the introduction into a population of some genetic factors, chromosomes or genes, from an outside source. Among bacteria, and possibly in other organisms although that is not yet established, this can occur in curious ways that introduce what is essentially a naked gene (a segment of a DNA molecule) into the cell's genetic system. Among many different organisms genetic migration occurs when an individual of one population breeds with one from another population: the familiar process of hybridization. The resulting hybrid, if viable and fertile, may then breed back into one population or the other and thus introduce genes from the foreign population. That may occur between distinct species, because often in plants and sometimes in animals the crossing of different species is merely relatively infrequent and not impossible. It occurs more frequently, indeed with considerable regularity, between adjacent local populations of the same species. Widespread species usually have local groups (demes) the individuals of which breed with each other more often than with those of other groups. The groups may, then, have somewhat different genetic pools, and the frequencies may be changed when the groups do interbreed.

The relationship of genetic migration to adaptation is not clear-cut. If hybridization produces fertile offspring, that in itself indicates a genetic compatibility that is not strongly inadaptive, at least. Hybrids may, indeed, be more vigorous than either parent. On the other hand, the mere fact of hybridization has no direct relationship to adaptation, and extensive migration of genes from one population to another is unlikely unless the spread is favored by natural selection.

Of the genetic processes so far discussed, recombination, mutation, and gene migration have important positive roles in maintaining or increasing variation within populations, and that is a prerequisite for evolutionary change. However, as regards the actual evolutionary changes they have no effect in themselves (recombination), or inadaptive or, so to speak, only accidentally adaptive effects (mutation, sampling error, gene migration). Yet it is abundantly clear, to say the least, that evolutionary change does occur and that it is usually adaptive. Evolution is not a random process and adaptation cannot be wholly, or indeed to any but a minimum extent, accidental. In essence, although not quite this clearly, Darwin had already found the answer to this riddle. The adaptive orientation of evolution must involve the one genetic process that is not necessarily or, as a matter of observation, usually, random: reproduction. Darwin himself glimpsed almost the whole of this process as well as was possible in pregenetic days, even though not always in clear detail. He did, however, tend to emphasize an unduly narrow concept of natural selection summed up as the survival of the fittest.

We now define natural selection as differential reproduction. The basic idea is a simple one. It is clear that in every population, from ameba to man and in all the rest, some individuals have more offspring than others, offspring that grow up and produce other offspring in their turn. If, now, the individuals that are thus more successful (or relatively prolific) in effective reproduction differ genetically, on an average and by however little, from the less successful individuals, their genetic characteristics will inevitably become more frequent in the genetic pool in the course of generations. Evolution will therefore occur, and it will be nonrandom, antichance, oriented in the direction of more successful reproduction. It has been demonstrated repeatedly, both in the laboratory and in the field, both by extensive theoretical models and by irrefutable observation, that this does occur. Direct demonstration for each and every one of the many millions of populations now in existence is of course impractical, and it is impossible for any of those that existed in the past. There is, however, a strong consensus that this is both a necessary and a sufficient explanation for all the nonrandom and adaptive phenomena observed in organisms living or fossil. It is the only nonrandom, inherently directional process known materially to occur in organic populations.

Simple as the core idea of natural selection is, its details in actual instances are extremely intricate and have not, indeed, been worked out in full or anything near it. For one thing, and at the most superficial level, relative success in reproduction can involve any or all of the stages in life cycles. In sexually reproducing or biparental organisms, which is to say most organisms of all kinds, male and female gametes (or the equivalent) must come together. In many animals, male and female individuals must be sexually acceptable and stimulating to each other and must mate. The gametes must fuse and must be genetically compatible. Normal embryological development must occur. The offspring must survive to breeding age. They must then themselves successfully breed and start the whole cycle over again. Selection depends on relative success in all these stages and at substages within them.

Darwin stressed the survival of offspring, and the Neo-Darwinians did so to even greater extent, considering selection primarily as differential mortality and hardly grasping the whole process as one of differential reproduction. Darwin also stressed the stage of mating in higher animals as sexual selection, which he distinguished from natural selection. Both differential mortality and sexual selection do, or at least can, lead to differential reproduction and hence they are simply special cases of the present broader concept of natural selection.

In considering just how selection acts in given cases, earlier selection theory erred in thinking of each trait or each gene (depending on whether one thought of hen-evolution or of egg-evolution) as a distinct thing. We have already noted that the genetic system of an individual *is* a system in which all the genes interact. Unless, then, a gene is quite exceptional in producing some one, very marked effect, it is hardly possible for selection to act on genes one by one. It acts on each in a broader context, or it acts on the combination of all genes. The genetic system of a population is also, in a different way, a single interacting system. All its genes and chromosomes, in their constant shuffling, must tend toward the most effective reproduction of the population as such, as a whole, and not necessarily just that most effective for one individual or another.

What one evolutionist has called good mixer genes will be favored. Those are the genes that tend toward normal development and vigorous offspring in combination with the other genes of the genetic pool. Genes that tend to annul or to modify the effects of disadvantageous mutations, as many do, will be selected for. There are also many genes that are more effective under given conditions if the paired chromosomes carry two different alleles of the gene rather than the same one, that is, if individuals are heterozygous rather than homozygous for the alleles. (A striking example is given in Chapter 14.) As a consequence of the Mendelian laws, the very first discoveries made in genetics, it is impossible for all the individuals of a population to be heterozygotic for any one gene. In such cases selection therefore produces an optimal balance between the more adaptive heterozygotes and the less adaptive homozygotes. Then, too, because each gene normally affects many characters and not just one, numerous genes have both adaptive and inadaptive effects. Selection then will not, because it cannot, act separately on the different effects, but will strike a balance between them at a point where advantage outweighs disadvantage.

Selection thus acts through the *whole* genetic systems of individuals on the *whole* genetic system of the population. If a particular combination of genes is especially advantageous, there are ways in which, in many instances, but not all, the combinations can be kept together and spread together through the population. They can be kept together on one chromosome and by changes in chromosome arrangement can be tied in a bundle, so to speak, with reduction or elimination of further recombination. Yet, for reasons already partly evident from the last paragraph, there is no one combination of genes uniform for all individuals that is at all likely to be optimal for the population as a whole. That concept of a single, ideal genetic pattern favored by selection in any one species was commonly held even a few years ago, but it has become naive.

There are other reasons why the optimum for a population is not a point, a single type either genetically or in bodily characteristics (either in egg or hen), but a range of variation in both respects. No population lives continuously in a uniform environment. A widespread species is spatially distributed over a varying landscape. Its common genetic pool must then provide for adaptive variation not, perhaps, ideal at any one place, but adequate in all. Through time, environments change from hour to hour, day to day, month to month, year to year, and so on up through the epochs, periods, and eras of geological history. Variation must be adequate to adapt to these changes as rapidly as they occur. There is here a balance between genetic variation, permitting adaptive changes by the genetic pool over, as a rule, longer periods, and breadth of genetically determined reaction ranges, permitting individual, non-hereditary adaptation over, usually, shorter periods. In almost all cases, genetic change is ultimately necessary if a lineage is to survive. A population with little or no genetic variation cannot adapt genetically: there are too few or no alternatives among which natural selection can select. Such a lineage will almost certainly become extinct sooner or later. And it is a fact demonstrated by paleontology that through the ages and for one reason or another the vast majority of species have become extinct.

What selection actually, directly favors is effectiveness of reproduction. That is almost a tautology, because we define natural selection as differential reproduction. But the major evolutionary problem that we set out to solve is adaptation to particular ways of life in particular environments. That is not the same thing as effective reproduction. These are, nevertheless, usual concomitants. As a rule, the animal best adjusted to its way of life in its environment will also in fact be most effective in reproduction. We cannot, however, assume that this is invariably so. Still less can we assume that selection will favor what, from our human point of view, is an upward progress. Myriads of examples of extinction and what we consider degeneration are there to contradict such facile assumptions. We can only say that when adaptation does occur and when there is progress, it is guided by natural selection, and not that those are necessarily the directions of selection. Here there are grave problems both philosophical and practical, some of which are discussed later in this book.

Now we can locate the feedback that we saw as necessary but that was vainly sought in Butler's wild speculations and in the Neo-Lamarckians' once legitimate but now disproven hypotheses. The feedback is an indirect, complex, usually slow, but ultimately extremely effective process occurring in populations through the course of generations: slightly (as a rule) in the passage from one generation to the next; greatly over a sequence of many generations. Its mechanism is natural selection. Its course is that genetic variation producing the most favorable relationship of organism to environment tends also to promote relative success in reproduction, which in turn spreads the favorable genetic variations, embodying new combinations and eventually new mutations of genes, through the populations in successive generations. The process cycles from genetics to organism-environment relationship to genetics again, and so round and round; from egg to hen to egg, or from hen to egg to hen, endlessly.

Here, too, we see the other essential of genetic coding that could never be read from chemical study of the DNA code itself: where the message came from. It was constructed by natural selection from what would otherwise be a meaningless jumble of code symbols. Perhaps I have a bias toward hen-evolution and organismal biology, but to me this is far the most important thing about the genetic message. Is it not more important to know what Shakespeare said and how he wrote his message than to analyze it into the letters of the alphabet or even the words of the dictionary?

There is another extremely important problem of evolution that has been only implicit up to this point. It was curiously neglected by Darwin, whose book called *The Origin of Species* is not really on that subject, but the neglect has been richly compensated in more recent years. The problem is how, in fact, *do* species originate; that is, not only how does one specific lineage evolve and adapt, but also how do multiple specific lineages arise and become divergently adapted. Only by such divergent and not only progressive adaptation have the waters, lands, and air been occupied by the swarming forms of life. Speciation in this sense of the multiplication of specific lineages has obviously been a constant and essential part of the history of life. It must have occurred many billions of times.

At first sight it could appear, and in fact did in Darwin's day, that natural selection solves this problem, too, and in quite a simple way. Let us take an extreme example for the sake of simplicity: a hypothetical species of omnivorous animals in which some individuals tend to eat more meat and some more vegetation. Why then would not natural selection tend to intensify those two different trends and perfect those adaptations until you wind up with one carnivorous and one herbivorous species, quite distinct from each other? The fatal flaw in such thinking becomes evident only when the situation is seen in more modern terms of populations and genetic pools. Our variant individuals are all drawing their genetic systems from the common pool of the species, and they are breeding with other individuals of the species. In this constant recombination and mixture, the majority of individuals will continue to be those least variant, in our example omnivorous and not carnivorous or herbivorous. The most carnivorous variants will necessarily breed as a rule with less carnivorous ones, and the most herbivorous with less herbivorous. Their offspring will not tend to diverge but to regress more or less toward the average condition—a fact that had been observed even before modern genetics was developed and that was raised as an objection to natural selection. Of course in such a population natural selection can effectively occur, but it will not divide the population into two. If the carnivorous variants regularly have more offspring than the herbivorous, the *whole* species as a single unit will evolve toward carnivorous adaptation. Similarly it could evolve toward herbivorous adaptation. It cannot, however, split into two populations evolving in different directions unless something stops the interbreeding among different variants in the population.

There has been much dispute as to whether there is any process that can limit or stop interbreeding among variants within the same local population, occupying the same area and in regular contact with each other. Everyone now agrees that this is highly unusual, and the continuing argument concerns only whether it *ever* occurs or not—a minor point that need not detain us here. Usually, then, speciation starts with the geographic separation of a segment of a specific population. Such separation in itself inhibits or prevents interbreeding. The separate local population is thus at least temporarily removed from the genetic pool of the rest of the species and in effect sets up a pool of its own. The separation is often temporary. Geographically separate local populations do get back in contact with each other and interbreed again, so that the species maintains its unity even though the pattern of its genetic descent is a network and not a single stream.

In many other instances, however, the separation of a local population endures long enough for it to strike out on its own, or in more measured terms for natural selection to have a divergent effect on it. If, for instance, our hypothetical omnivorous species had two geographically separate populations, one in a region with rich animal food and little competition for it and the other in a region similarly favoring vegetarianism, then, indeed, selection could produce separate carnivorous and herbivorous species from the single ancestral species. The genetic pools and the other characteristics of separate segments of what was once one population will inevitably diverge in time. Eventually there will come a time when they will not resume interbreeding even if they come in contact with each other. Then the genetic separation has become irreversible and full speciation has occurred.

Anything that strongly limits or prevents interbreeding between species, or in other words that promotes reproductive isolation, is called an isolating mechanism. Such mechanisms take many different forms, some quite strange. Two species of animals may breed at different seasons, or even at different times of day. Many animals have sexual preferences, which may be associated with distinctive color patterns or other recognition marks. Many go through more or less elaborate courtship rituals characteristic of a species and essential for successful mating. Pollinating insects tend to concentrate on single species of plants and hence can be isolating mechanisms for those plants. Sooner or later as species diverge their genetic systems become so different as to be incompatible, that is, the combination of gametes from male and female parents can no longer produce normal offspring. They may produce vigorous but sterile offspring, like mules, or offspring with reduced vigor or fertility, or simply no offspring at all. The ability of species to hybridize, even to the production of quite abnormal offspring, implies rather close relationship. Genetic divergence of distantly related animals and plants has made interbreeding flatly impossible.

Even this brief and superficial account of one modern approach to evolution shows that the whole process is exceedingly complex, perhaps bewilderingly so. Recombination, mutation, sampling error, and gene migration in genetic pools. With those, the cycling of genetic system through individual development to organism-environment interaction to differential reproduction and hence feedback into the genetic system and on to endless recycling—in short, natural selection. Adaptation as a somewhat indirect outcome of that cycle, but too closely linked and too essential to be considered a by-product. Geographic separation, reproductive isolation, and isolating mechanisms as keys to the vast proliferation of species. All those and other factors not even mentioned enter into the evolution of life. Consider what is to be explained, what we thus seek to understand: the boundless exuberance of life manifest in millions of species; their rise and change through billions of years; the delicate coordination within each organism; the awesome intricacy of each; the pervasive adaptation and balance of nature. No wonder that the primitive mind took the short cut of ascribing all this to miracle. No wonder that our long way around of learning how the miracle was really wrought proves to be far from simple, and yet beautiful in its complexity.

PART TWO

Evolution Among the Sciences

CHAPTER FIVE

Biology and the Nature of Science

EVOLUTION is only one aspect, although the most pervasive aspect, of biology. Biology is only one, albeit the most unifying one, of the sciences. For that matter, science is only one of many human activities, but it is at least among the most important of them. Let us consider the role of evolution in the explanations of biological phenomena, the role of biology among all of science, and the significance of science itself. For a start, what in fact *is* science?

There is a whole library of attempts to define science. The literature is so prolix and in part so contradictory that I cannot analyze it and should perhaps hesitate to add to it. The element of confusion is well illustrated by a recent statement that science is "thinking about the world in the Greek way." That is in fact an important thing that science is *not*. It has often been argued that the Greek sense of order was a necessary condition for the rise of science. Necessary, perhaps; sufficient, definitely no. The actual origin of science in the modern sense involved a revolt against thinking in the Greek way.

The Greek way, which became traditional in medieval Europe, was well expressed by Plato, for example, when he said in *The Republic*, "We shall let the heavenly bodies alone, if it is our design to become really acquainted with astronomy." In other words the essence of things was believed to reside in a philosophical ideal, and observation of real phenomena was considered not only unnecessary but also positively wrong. Some five centuries after Plato, Ptolemy again formalized the Greek way and helped to imbed it in Western thought for another 1500 years when he said that the goal of astronomy was "to demonstrate that all heavenly phenomena are produced by uniform circular motion." Now, that is not physically true, and Ptolemy knew that it was not. He was explicit that his intention was not to explore physical reality. The early astronomers' only gestures toward reality were attempts to "save the appearances," that is, to try to eliminate obvious contradictions without abandoning their a priori philosophical ideals, such as that of uniform circular motion. "Saving the appearances" was a euphemism for saving the philosophical postulates. Facts were not to be explained, but to be explained away.

Science was born when a few thinkers decided that appearances were something not to be saved but to be respected. Those hardy souls—Copernicus, Galileo, and Kepler were among them—eventually abandoned the Greek way of deciding how things ought to be and gave us our way of observing how in fact things *are*. Definitions of science may differ in other respects, but to have any validity they must include this point: the basis of science is observation. This may be expressed and applied in different ways. Francis Bacon, a close contemporary of Kepler, had a concept of science as gathering all possible observations and then deriving from them generalizations and laws by induction, in accordance with an elaborate system rooted in Scholastic logic. It has often been pointed out that Bacon's system does not really work and has not been followed by any successful scientist (Bacon himself was not one). Nevertheless his respect for observation and his operational approach to science are among the points in which his influence has been profound and beneficial.

Thinking of science in terms of methods came more and more into vogue with the scientific triumphs of the nineteenth century and science's great acceleration into the twentieth. Indeed one still, although now less frequently, hears of teaching *the* scientific method, as if science were a set routine applicable to any subject. That tendency reached a climax with Karl Pearson and others a generation or so ago. The scientific method was sometimes formalized as involving six successive operations:

1. A problem is stated.
2. Observations relevant to the problem are collected.
3. A hypothetical solution of the problem consistent with the observations is formulated.
4. Predictions of other observable phenomena are deduced from the hypothesis.
5. Occurrence or nonoccurrence of the predicted phenomena is observed.
6. The hypothesis is accepted, modified, or rejected in accordance with the degree of fulfillment of the predictions.

There is no question that such a cut-and-dried method does work in particular instances, or that each of the six operations is essential in various phases of scientific research. Nevertheless, the formulation fails as an over-all characterization of science. It is not a definition; it says nothing of the goals or the nature of science. Its implication of a general routine that automatically solves any scientific problem is false. It quite ignores the most difficult, most creative, and most important elements of scientific endeavor. How does one discern a problem, or decide what kinds of questions are to be asked? How does one determine what observations are relevant? And especially, what kind of hypothetical solutions are acceptable and where do they come from? Perhaps the most cogent objection of all is that important basic research has seldom really followed the "method" just as it is stated.

In our own days James Conant has strongly criticized that kind of formulation and has proposed a new definition of science and another characterization of its methods. His definition is: "An interconnected series of concepts and conceptual schemes that have developed as a result of experimentation and observation and are fruitful of further experimentation and observation." He characterizes scientific method as comprising: "(1) speculative general ideas, (2) deductive reasoning, and (3) experimentation." Like all brief statements on any subject, these are ambiguous and incomplete outside of the expanded context given them by the author. The definition, taken by itself, does not define. If reread, it will be found to apply perfectly to the work of Picasso, for example, and although Picasso's work is certainly creative and great it is with equal certainty not science. Of course Conant's point is to emphasize the dynamic, ongoing nature of science. That is a characteristic of the most important scientific investigations, but dynamism is not confined to science and does not characterize all of science.

Conant may also be too hard on his predecessors. His summary of scientific method is freer, more impressionistic, than the earlier formulation and to that extent more nearly covers the varied gambits of research. However, in some respects it is less complete or explicit. It starts with the formulation of the hypothesis, the third step of the earlier summary, and its other two steps are essentially the fourth and fifth. The sixth step is not eliminated but is simply taken for granted and not stated.

The main virtue of Conant's formulation is its recognition of the role of speculation, intuition, or just plain hunch in finding a hypothesis. It ignores the fact that some observation inevitably precedes the speculation, and both formulations fail to note that observation (whether of nature or of an experiment) always is the first step in any scientific investigation. No one ever had a hunch that was not *about* something—in the case of science, about possible relationships among facts already known.

Conant makes the essential point that the aim of science is to seek and verify general ideas, relationships, and interconnections among phenomena. Obviously science has nothing to do and cannot exist if phenomena have not, in fact, been observed, but there science begins, not ends. It follows that although the observation of facts or memorization of data is a necessary basis and accomplishment of science, that in itself is not science. Science, truly to be such, must center not on descriptions and names but on principles—that is, generalizations, theories, relationships, interconnections, explanations about and among the facts.

A second point often left implicit but requiring meticulous

attention is that the materials of science are literally material. The observations of science are of material, physically or objectively observable phenomena. Its relationships are material, natural relationships. This is not to say that science necessarily denies the existence of nonmaterial or supernatural relationships, but only that, whether or not they exist, they are not the business of science. This requires, if you like, a measure of self-discipline among scientists, a recognition that their methods do not work properly in the absence of this restriction.

The third feature that distinguishes science from other fields of thought and activity is that it is self-testing by the same kinds of observations from which it arises and to which it applies. It is, to use a currently popular but perhaps overworked bit of jargon, a cybernetic system with a feedback that in spite of oscillations keeps its orientation as nearly as may be toward reality. That is the point of the deductive phase stressed by both Pearson and Conant as well as by almost all other modern writers on scientific method, although in fact formal deduction is not invariably involved in scientific self-correction. On that more must be said, but here we have reached a point where another attempt to define science is in order.

One way to approach definition is to consider science as a process of questioning and answering. The questions are, by definition, scientific if they are about relationships among observed phenomena. The proposed answers must, again by definition, be in natural terms and testable in some material way. On that basis, a definition of science as a whole would be. Science is an exploration of the material universe that seeks natural, orderly relationships among observed phenomena and that is self-testing. We may well add, but not as part of a definition, that the best answers are theories that apply to a wide range of phenomena, that are subject to extensive tests, and that are suggestive of further questions. It is such theories that contribute most to the ongoing aspect of science so properly stressed by Conant. Nevertheless, most scientific endeavor has more limited objectives, and some endeavor, even though scientific by definition, has no evident sequel.

It is noteworthy that almost all studies of the philosophy and methods of science have referred primarily to the physical sciences. That is in part because the physical sciences do have a primacy— not, I insist, logically, but historically. The first sciences, as we now strictly define science, were physical sciences. That was at a time when scientists considered themselves to be also, or even primarily, philosophers, and indeed "natural philosophy" was long synonymous with "physics." The tradition has persisted. It has been reinforced by the reductionist half-truth (of which more later) that all phenomena are ultimately explicable in strictly physical terms. Another factor has been the prestige accruing from the thorough and more obvious impingement of the physical sciences on daily life through technology. It is also possible that more of the most brilliant and thoughtful minds have gone into the physical sciences; I prefer not to think so, but I suspect there is some truth in that.

The point here is that most considerations of the history, methods, and nature of science have been heavily biased by concentration on physical science and not on science as a whole. That has been notably true of concepts of scientific laws, of predictability, of the testing of hypotheses, and of causality. Francis Bacon warned, "Though there are many things in nature which are singular and unmatched, yet it [the human understanding] devises for them parallels and conjugates and relatives which do not exist." Nineteenth-century physicists did not heed his warning. They refused to consider the unique object or event and assumed that all phenomena could be reduced to supposedly invariable and universal laws such as the gas laws or the law of gravitation. It followed that, once a law was known, its consequences could be fully predicted. The consequences deduced from a hypothesis became predictions as to what would happen if an experiment were performed, and that is the pertinent test embodied in Pearson's, and still in Conant's (and many others'), descriptions of scientific method. It further followed —or the physical scientists thought it did—that when a law successfully predicted an event, the law explained the event as a result and specified its causes.

Here we in the twentieth century have seen something curious and indeed almost comic happen. Physicists have found that some, at least, of their laws are not invariable; that their predictions are statistical and not precise; that some observations cannot in fact be made; and that absolute confirmation by testing of a hypothesis therefore cannot be obtained. Many have gone further and concluded that causality is meaningless and even that order in nature—the last *scientific* relic of our Greek heritage—has disappeared. That is, of course, the so-called scientific revolution wrought by quantum theory and the principle of indeterminacy. The physicists' reactions to this (even in my very limited knowledge of them) ran the gamut from reason to hysteria. Some—Bridgman is a sad example—found science coming apart in their hands, further scientific knowledge impossible, and the universe and existence itself left wholly meaningless. Others, such as Jeans, also accepted the whole idea of orderlessness and acausality but, with almost mystical glee, likened the release from physical law to release from prison. Still others have had what seems both the most mature and the most scientific reaction: they have concluded that the physicists have failed somewhere and that there must be some rational way to get over the difficulty.

The aspect that I spoke of as almost comic is this: well before the "revolution" life scientists had observed that laws, in the rigid nineteenth-century conception of physics, do not apply to many phenomena in nature. Further, they know that prediction (not the only way of testing hypotheses) is commonly statistical and no less scientific or confirmatory of a hypothesis for all that. They knew that this is no contradiction of the orderliness of nature, and they discerned that only an unnecessarily restricted concept of causality is affected. The "revolution" was a revolution only for those who had insisted that everything must be explained ultimately in terms of classical physics—and where were there ever any real grounds for such a narrow view of science? It is true that understanding of statistical law and polymodal causality had crept over the life scientists gradually, so that the impact of these concepts was not seen as revolutionary. It is also true that not many biologists are given to exploring the philosophical implications of their science. There was therefore little really clear discussion of causality in biology before that by Ernst Mayr in 1961.

A fundamental, though not a sufficient, criterion of the self-testability of science is repeatability. Norman Campbell's definition of science as "the study of those judgments concerning which universal agreement can be obtained" emphasizes this point. That is indeed not so much a definition of science as of its field and its connection with reality. Campbell's meaning is that the data of science are observations that can be repeated by any normal person. That is as true of, say, the observation of a fossil tooth under a microscope as it is of the height of mercury in a tube in Torricelli's famous experiment, or of more recent observations of protein separation by chromatography and electrophoresis. Illusion, even to the point of hallucination, is always a possibility, but it is one that can be eliminated for all practical purposes by repetition of observations, especially by different observers and different methods. It is also true that unique events occur, but evidence on them is acceptable if there is confidence that anyone else in a position to observe them would have observed them.

In what used to be called the exact sciences, which have turned out not to be so exact, it was formerly assumed that uniform phenomena had absolute constants measurable to any degree of accuracy. As a very simple example, the length and period of a pendulum were assumed to have infinitely exact and determinable values. It now appears that this is not necessarily true, and that is one of the discoveries that so upset the physical scientists. But in the actual practice of observation it has always been evident that infinitely exact measurement is impossible. All that repetition and instrumental refinement can do is to generate a degree of confidence that a measurement (at any given time and under given conditions) lies within a certain range. Inference from the observation takes into account the size of the range and the degree of confidence. The conclusion that even in principle the range cannot be infinitely small and confidence infinitely great makes no difference operationally, at least.

It is further true that with many phenomena the whole point of observation is not an exact measurement or determination of occurrence but establishment (again to some degree of confidence) of a probability. The classical example is the tossing of a coin, and here the biologists' point is that we do *not* expect the probability of throwing heads to be exactly one-half. As modern scientists and not ancient Greeks, we are examining real, objective coins and not the Platonic idea of a coin. By repeated observation of a real coin, we can establish a high degree of confidence that the probability is in a certain range. If the range is large, it is likely to include the probability of one-half, but if the range is made small it is likely to exclude that a priori ideal. Analogous phenomena are very common in biology. For example, we do not expect an expanding population of flies to spread according to an exact law. We expect only to achieve confidence that the rate will be within a certain range of probability, or to construct a frequency distribution of rates. Discovery that Boyle's "law" has the same probabilistic nature neither surprises nor upsets us. We would expect it, because the molecules of gas, like the flies, are real individuals which, however alike they are in other respects, have had different histories. The Greeks could, but a scientist cannot, be concerned with the ideal gas of classical physics. Perhaps the revolution in physics was only the final severing of the umbilical cord from ancient Greece.

The most widespread and conclusive process of self-testing in science is testing by multiplication of relevant observations. In the natural sciences it is impossible to prove anything in the absolute sense of, for example, a proof in mathematics. Multiplication of observations can only increase our confidence within a narrowing range of probability. If confidence becomes sufficiently great and the range is encompassed by the hypothesis, we begin to call the hypothesis a theory, and we accept it and go on from there. The test is, of course, whether the range of probability is in fact within the scope of the hypothesis—in other words, whether the observations are consistent with the hypothesis.

A key word in the expression "multiplication of relevant observations" is *relevant*. The simplest definition is that relevant observations are those that could *disprove* the hypothesis, for disproof is often possible even though absolute proof is not. The more observations fail to disprove a hypothesis, the greater the confidence in it.

Prediction in the classical sense is a special case of that general procedure. From the hypothesis consequences are deduced such that their failure to occur would disprove the hypothesis. Of course their occurrence would not *prove* anything; it would only increase confidence. That this is in fact a special case and not the touchstone of scientific theory is easy to demonstrate. Again, examples are more familiar to biological than to physical scientists, although they occur in both fields. The most striking example is the most important of all biological theories: that of organic evolution. Although some quite limited predictions can be deduced from the theory, the theory was not in fact established by prediction and is not sufficiently tested by it. An enormous number of observations enormously varied in kind are all consistent with this theory, and many of them are consistent with no other theory that has been proposed. We therefore can and, if we are rational, must have an extremely high degree of confidence in the theory—higher than legitimate confidence in many things we call "facts" in daily life. That kind of nonpredictive testing most commonly occurs in fields that have a temporal or historical element, such as evolution among the biological sciences or the time-linked processes in geology among the physical sciences. In fact a neglected historical component also affects many physical laws, as in the example of the histories of the individual molecules in a gas.

In discussing the nature and basic procedures of science I have been quite free in using such expressions as "reality," "phenomena," "the material universe," and so on. Philosophers have long since pointed out, and philosophical scientists are still worried by, the fact that the very existence of such a thing as objective reality is uncertain. I have already referred to Bridgman's despairing conclusion that "the very concept of existence becomes meaningless." In *The Scientific Outlook* Bertrand Russell has discussed this matter more optimistically if equally inconclusively. He points out (in more and different words) that what we call observing a phenomenon is in fact only sensing certain events that occur to and within ourselves. For example, when we think we have seen something, we know only the event that light quanta of certain energies and patterns impinged on our retinas and produced other events in our nervous system. The object we think we saw "remains veiled in mystery." Russell asks finally, "Are circumstances ever such as to enable us, from a set of known events [for example, those in our nervous system] to infer that some other event [for example, the material existence of what we think we see] has occurred, is occurring, or will occur?" He concludes, "I do not know of any clear answer. . . . Until an answer is forthcoming, one way or another, the question must remain an open one, and our faith in the external world must be merely animal faith."

Now, some feel that this is nonsense and that sensible people will not waste time on it. Whether or not there *really* is an external world, we certainly have to act as if there were, so we may as well ignore the question. Indeed I shall not here spend much time on it, but it has bothered many scientists, so it does seem worth while to point out that there *is* an answer. In fact there are several. Russell himself has provided one, apparently unwittingly, although it is dangerous to assume that he is ever unwitting. His example of what he calls the "known events" includes light from the sun bouncing off a man named Jones and then entering the eye. "Jones himself" may still be "wrapped in mystery," as Russell says, but evidently *something* happened out there. The faith required is not that "out there" exists, but that what happens "in here" contains some information about it. Such an answer obviously does not supply a philosophical absolute, but it should satisfy a scientist's more modest demand for reasonable confidence.

Norman Campbell has pointed out that the fact that others demonstrably receive the same sensations as we do from the same stimuli is evidence that the outer world does exist. That is the basis for his remark, quoted earlier, on the obtainability of universal agreement in [observational aspects of] science. It is also evidence that the stimuli are structured—that is, do convey information. Again a philosopher may quibble and say that the reactions of others have no bearing if the others are not really there, but a scientist will gain another degree of confidence.

Still another consideration seems to me the most interesting of all. It is, in a sense, a validation of the "animal faith" given by Russell (after Santayana) in the passage quoted earlier, as sole basis for assuming that we really can obtain knowledge of the outer world. The fact is that man originated by a slow process of evolution guided by natural selection. At every stage in this long progression our ancestors necessarily had adaptive reactions to the world around them. As behavior and sense organs became more complex, perception of sensations from those organs obviously maintained a realistic relationship to the environment. To put it crudely but graphically, the monkey who did not have a realistic perception of the tree branch he jumped for was soon a dead monkey—and therefore did not become one of our ancestors. Our perceptions do give true, even though not complete, representations of the outer world because that was and is a biological necessity, built into us by natural selection. If it were not so, we would not be here! We do now reach perceptions for which our ancestors had no need, for example of x-rays or electrical potentials, but we do so by translating them into modalities that are evolution-tested.

That is one of the several senses in which science itself, as a whole, is fundamentally biological. A second sense in which that is true is involved in another point that has lately been bothering the physicists. The point is that whenever a scientist observes anything he is himself part of the system in which the observing takes place. He therefore should not assume that what he observes would be exactly the same if he were not observing it. But he cannot very well observe what happens when he is not observing! Therefore, the argument runs (but personally I do not run with it), there is no such thing as objective knowledge, and the goals of science are wholly delusive. Some atomic physicists say this does not matter as far as the man-sized world is concerned but matters only when you get down to their invisible, but all too obviously not imaginary, objects of study. Yet I really do not see why size matters in principle. In either case the system actually observed contains something alive—to wit (as a minimum), the observer. Surely it would never occur to anyone but an atomic physicist that because a system includes something alive it cannot be properly studied!

To suppose that study, to be objective, should exclude the observer is as unrealistic as Plato. Science is *man's* exploration of his universe, and to exclude himself even in principle is certainly not objective realism—unless you insist that his inclusion is subjective by definition, but that would be merely playing with words. And to say that we cannot learn anything materially factual about a situation if we ourselves are in it is utter and nonsensical negation of the very meaning of learning. The essential in objectivity is not the pretense of eliminating ourselves from a situation in which we are objectively present. It is that the situation should not be interpreted in terms of ourselves but that our roles should be interpreted realistically in terms of the situation. To a biologist the discovery (to call it such) that every system observed includes the observer has quite a simple meaning. It merely means that all systems in science have a biological component.

There is another, related sense in which all science is partly biological. It is all carried on by human beings, a species of animal. It is in fact a part of animal behavior, and an increasingly important part of the species-specific behavior of *Homo sapiens*. From the functional point of view, it is a means of adapting to the environment. It is now, especially through its operating arm, technology, the principal means of biological adaptation for civilized man. It is an evolutionary specialization that arose from more primitive, prescientific means of cultural adaptation, which in turn had arisen from still more primitive, prehuman behavioral adaptation. I recently had occasion to point out to some ethnologists that culture in general is biological adaptation and that they could resolve some of their squabbles and find the common theoretical basis that eludes them if they would just study culture from this point of view. The suggestion was not well received, but it is true just the same. Some thought I was being a racist and some thought I was being a Social Darwinian, both quite rightly pejorative epithets in ethnological circles. Of course I was being neither one. I was just being a biologist drawing attention to the really quite obvious fact that culture is a *biological* phenomenon. That is true, in heightened degree, of the special aspect of culture we call science.

As Gillispie has admirably shown in his book *The Edge of Ob-*

jectivity, the rise of science, in the strictest modern sense of the word, centered on increasing insistence on objectivity. It now seems clear that in some instances that insistence went too far. I have noted that some scientists reached the unnecessary and, in the last analysis, absurd position that complete objectivity would exclude the observer. Since exclusion of the observer is obviously impossible in the practice of science, scientists who held that view, as we have seen, tended either to fall into despair or to revert to various more or less covert forms of idealism. I have here maintained that this was an unnecessary casualty and that the concept of objectivity essential to science is saved by recognition that scientific objectivity has a biological component. A related casualty that was almost inevitable in the struggle to develop modern science involves the concept of teleology.

The doctrine of final cause, of the end's determining the means, was another essential element in Greek thought, which was anthropomorphic in a truly primitive way. This doctrine was probably an inevitable outcome of introspective and deductive philosophy. Rational human actions are largely explicable by their purpose, by the results they are expected to produce. It therefore seemed logical to conclude that the orderly intricacy of the world at large was in a similar way purposeful and governed by a foreseen end. Such concepts were particularly important to Aristotle, and through his works they came to be held as almost axiomatic in the western European milieu in which science finally arose. The broadly philosophical position was that things exist, or events occur, as prerequisites of their results, and that the result, as final cause, is the real principle of explanation. In more popular form, this view led to the belief that nature exists only for and in relation to man, considered as the ultimate purpose of creation or the overriding final cause.

As physical science became more objective, it was apparent that teleology, even if not rejected as a philosophy, had to be ignored as a means of scientific explanation. The scientist, as such, asked "What?" or "How?" about phenomena such as gravity or gas pressure, not "Why?" or "What for?" Description of how things fall, in terms of masses, distances, and gravitational constants, is clearly scientific, but the question, "What do things fall for?" seems unscientific. It elicits no objectively testable answers. It was thus inevitable that the strictest scientific attitude should endeavor to exclude any form of teleology, and in the physical sciences there seemed to be no great difficulty in excluding it. One could, at least, readily evade teleology by ascribing physical laws to a first rather than a final cause, although even here the usual philosophical or theological belief continued to be that natural laws exist in order to make the world a suitable habitat for man.

In the biological sciences the elimination or even the brushing aside of crude teleology was incomparably more difficult, and that is a principal reason why a fully scientific biology lagged so far behind a scientific physics. It is not necessary or perhaps even possible to see any immediate, inherent purpose in a stone's falling, but it is quite inevitable that an animal's seeking its food should be interpreted in terms of purpose or, at least, of an end served. All organisms are clearly adapted to live where and how they in fact live, and adapted in the most extraordinary, thoroughgoing, and complex ways. In fact they plainly have the adaptations in order to live as they do. The question, then, is how those key words *in order to* are to be interpreted. Until a century or so ago it occurred to very few naturalists to interpret them in any but the classical teleological way. For example, to Cuvier, high priest of natural history in the early nineteenth century, the validity of fully Aristotelian teleology seemed self-evident, and it was the heart of his theoretical system. Cuvier went all the way to a man-centered teleological conception of the universe. He could think of no better reason for the existence of fishes—which he considered poor things, even to the watery, unromantic nature of their *amours*—than that they provided food for man. That was also the period in England of Paley's *Natural Theology* and, later, of the *Bridgewater Treatises* "on the power, wisdom, and goodness of God, as manifested in the creation"—that is to say, on Christian teleology as a necessary and sufficient explanation of nature, and most particularly of animate nature.

The facts of adaptation *are* facts, and the purposeful aspect of organisms is incontrovertible. Even if the explanation offered by

Aristotelian, and much later by what was then orthodox Christian, teleology were true, that would definitely be an article of faith and not of objectively testable science. Thus it was necessary either to conclude that there is no scientific explanation of organic adaptation or to provide an acceptable, testable hypothesis that was scientific. Before Darwin most biologists accepted the first alternative, which (although few of them realized this fact) meant quite simply that there could be no such thing as a fully scientific biology. It was Darwin, more than any other one person, who supplied the second alternative. In *The Origin of Species* he made no entirely clear distinction between establishing the fact that evolution has occurred and proposing a theory as to how natural processes could produce organic adaptation. He has therefore been accused of unnecessarily confusing two issues that should have been kept quite separate, but that was not really the case. Evolution itself becomes a nonscientific issue if the explanation of adaptation in the course of evolution is left in the field of metaphysics, philosophy, and theology. Darwin really went to the heart of the matter with unerring insight. Explanation of adaptation was the key point, and Darwin demonstrated, at the very least, that a natural, objective explanation of adaptation is a rational possibility and a legitimate scientific goal. That, at long last, made biology a true and complete science.

Darwin fully respected the appearances and made no attempt to save them by explaining them away. The hand of man, for example, *is* made for grasping. Darwin said so, and then provided a natural scientific explanation for the fact. He thus did not ignore the teleological aspects of nature but brought them into the domain of science. Some of Darwin's contemporaries and immediate successors recognized that fact by redefining teleology as the study of adaptation and by pointing out that Darwin had substituted a scientific teleology for a philosophical or theological one. The redefinition did not take. The older meanings of the word *teleology* were ineradicable, and they brought a certain scientific (although not necessarily philosophical) disrepute to the whole subject.

The physical scientists had earlier, and more completely, evaded the issues of classical teleology. By the end of the nineteenth century, if not before, it had become for them virtually a dogma that a scientist simply *must not* ask, "What for?" Physical scientists considered the question as applied to natural phenomena either completely meaningless or, at best, unanswerable in scientific terms. Such was the priority and primacy of the physical sciences that this position even came to be widely considered a necessary qualification of truly scientific endeavor, part of the definition of science. That led in turn to a very curious development that was at its height in the 1920's and is still exerting a strong but now more clandestine effect. Many biologists threw out the baby with the bath water. In seeking to get rid of nonscientific teleology they decided to throw out all the quite real and scientific problems that teleology had attempted to solve.

That took several different forms. One form in evolutionary studies was the mutationist belief that organisms do not become adapted to a way of life but simply adopt the way of life that their characteristics, originating at random, make possible. Another form was behaviorism, which also, in essence, sought to eliminate adaptation as a scientific problem by refusing to consider behavior as motivated, as goal-directed, or even as serving needs (and hence in some sense having purpose) in the organism as a whole. Behaviorism strove to be primarily descriptive, and what explanatory element was admitted was meant to be confined to consideration of the physiological substrates and concomitants of the behavior described. It is that latter aspect that still influences a considerable segment of opinion in biology, confining biological explanation to the physicist's question "How?" and eschewing "What for?" This attitude, still strongly held in some quarters, involves the idea that scientific explanation must be reductionist, reducing all phenomena ultimately to the physical and the chemical. In application to biology, that leads to the quite extraordinary proposition that living organisms should be studied as test-tube reactions and that their being alive should enter into the matter as little as possible. As behaviorism omits the psyche from psychology, so this form of reductionism omits the bios from biology.

Those tendencies were unquestionably salutary in some respects. They have helped to eliminate the last vestiges of pre-Darwinian teleology from biology. They have also helped to counteract vitalistic, metaphysical, and mystical ideas which, whatever one may think of them in their own sphere, are completely stultifying as principles of scientific explanation. Here, however, the reductionist tendency has been two-edged. By seeming to negate the very possibility of scientific explanation of purposive aspects of life, it has encouraged some biologists, who insist that such aspects nevertheless exist, to seek explanations quite outside the legitimate field of science. Naming of names is perhaps invidious, but to show that I am here setting up no straw man I will just mention Teilhard de Chardin in Europe and Sinnott in the United States. (See Chapter 11.)

The reaction went much too far. It went so far as to falsify the very nature of biology and of science through supine acceptance of a dictum that all science is in essence physical science. In fact, the life sciences are not only much more complicated than the physical sciences, they are also much broader in significance, and they penetrate much farther into the exploration of the universe that *is* science than do the physical sciences. They require and embrace the data and *all* the explanatory principles of the physical sciences and then go far beyond that to embody many other data and additional explanatory principles that are no less—that are, in a sense, even more—scientific.

This can be expressed, as Mayr, Pittendrigh, and others have expressed it, in terms of kinds of scientific explanations and kinds of questions that elicit them. "How?" is the typical question in the physical sciences. There it is often the only meaningful or allowable one. It must also always be asked in biology, and the answers can often be put in terms of the physical sciences. That is one kind of scientific explanation, a reductionist one as applied to biological problems: "How is heredity transmitted?" "How do muscles contract?" and so on through the whole enormous gamut of modern biophysics and biochemistry. But biology can and must go on from there. Here, "What for?"—the dreadful teleological question—not only is legitimate but also must eventually be asked about every vital phenomenon. In organisms, but not (in the same sense) in any nonliving matter, adaptation *does* occur. Heredity and muscle contraction do serve functions that are *useful* to organisms. They are not explained, in this aspect, by such answers to "How?" as that heredity is transmitted by DNA or that energy is released in the Krebs cycle.

In biology, then, a second kind of explanation must be added to the first or reductionist explanation made in terms of physical, chemical, and mechanical principles. This second form of explanation, which can be called compositionist in contrast with reductionist, is in terms of the adaptive usefulness of structures and processes to the whole organism and to the species of which it is a part, and still further, in terms of ecological function in the communities in which the species occurs. It is still scientifically meaningful to say that, for instance, a lion has its thoroughgoing adaptations to predation *because* they maintain the life of the lion, the continuity of its species, and the economy of its communities.

Such statements exclude the grosser, man-centered forms of teleology, but they still do not necessarily exclude a more impersonal philosophical teleology. A further question is necessary: "How does the lion happen to have these adaptive characteristics?" or, more generally and more colloquially, "How come?" This is another question that is usually inappropriate and does not necessarily elicit scientific answers as regards strictly physical phenomena. In biology it is both appropriate and necessary, and Darwin showed that it can here elicit truly scientific answers, which embody those that go before. The fact that the lion's characteristics are adaptive for lions has caused them to be favored by natural selection, and this in turn has caused them to be embodied in the DNA code of lion heredity. That statement, which of course summarizes a large body of more detailed information and principle, combines answers to all three questions: not only "How?" and "What for?" but also "How come?" Always in biology but not invariably in the physical sciences, a full explanation ultimately involves a historical—that is, an evolutionary—factor.

Here I should briefly clarify a point of possible confusion. Insistence that the study of organisms requires principles additional to those of the physical sciences does not imply a dualistic or vitalistic view of nature. Life, or the particular manifestation of it that we call mind, is not thereby necessarily considered as nonphysical or nonmaterial. It is just that living things have been affected for upward of two billion years by historical processes that are in themselves completely material but that do not affect nonliving matter, or at least do not affect it in the same way. Matter that was affected by these processes became, for that reason, living, and matter not so affected remained nonliving. The results of those processes are systems different in kind from any nonliving systems and almost incomparably more complicated. They are not for that reason necessarily any less material or less physical in nature. The point is that *all* known material processes and explanatory principles apply to organisms, while only a limited number of them apply to nonliving systems. And that leads to another point, my final one.

When science was arising, Francis Bacon insisted that all its branches should be incorporated into one body of fundamental knowledge. Bacon placed this in an Aristotelian framework really inappropriate for modern science; he wrote at a time when one mind could grasp the essentials, at least, of all the sciences; and he was not himself a practicing scientist. Of course nowadays, as regards detailed knowledge and adequate research ability, there is no such thing as a general scientist, a general biologist, or even a general entomologist. In the practice and teaching of science, specialization and the accompanying fragmentation of the sciences have become absolutely necessary. Yet this practical necessity has not eliminated the force and value of the conception that the universe and all its individual phenomena form one grand unit and that there is such a thing as science, not just a great number of special and separate sciences.

Bacon further maintained that the unity of nature would be demonstrated and the sciences would be incorporated into one general body by a fundamental doctrine, a *Prima Philosophia,* uniting what is common to all the sciences. Despite the great change in philosophical outlook, that has become a traditional approach to the unification of the sciences. In our own days, Einstein and others have sought unification of scientific concepts in the form of principles of increasing generality. The goal is a connected body of theory that might ultimately be *completely* general in the sense of applying to *all* material phenomena.

The goal is certainly a worthy one, and the search for it has been fruitful. Nevertheless, the tendency to think of it as *the* goal of science or *the* basis for unification of the sciences has been unfortunate. It is essentially a search for a least common denominator in science. It necessarily and purposely omits much the greatest part of science, hence can only falsify the nature of science and can hardly be the best basis for unifying the sciences. I suggest that both the characterization of science as a whole and the unification of the various sciencies can be most meaningfully sought in quite the opposite direction, not through principles that apply to all phenomena but through phenomena to which all principles apply. Even in this necessarily summary discussion, I have, I believe, sufficiently indicated what those latter phenomena are: they are the phenomena of life.

Biology, then, is the science that stands at the center of all science. It is the science most directly aimed at science's major goal and most definitive of that goal. And it is here, in the field where all the principles of all the sciences are embodied, that science can truly become unified.

CHAPTER SIX

The Study of Organisms

The goal of biophysics is to formulate an equation and state the boundary conditions of a given living system. The working out of the equation under the given conditions would predict the future behavior of the organism.

Biology starts with biochemistry and goes on to neurophysiology and genetics. All else is stamp collecting.

Genetics, the central discipline of biology, has as its ultimate subject a group of chemical compounds. Genetics requires knowledge of the physical sciences but no other knowledge in the field of biology itself.

Biology is the study of organisms. Physical and chemical research on organic materials has no biological significance unless it is related to whole organisms and populations of organisms.

IN the last chapter we saw biology in a unifying role among the sciences. Yet that central role has introduced some disruptive tendencies within biology itself. This science touches all others and can be approached from many different directions. Enthusiasts for one point of view or another sometimes overlook the total organization and essential unity of biology. An illustration of such natural but unfortunate exaggeration is given by the four statements above, which are based on specific sources but are paraphrased without citation of the original authors. They reflect widespread attitudes, and that, not ascription to one biologist or another, is the important point.

Of course I have a bias, too, and mine is toward the last of those four biased statements. However, I am not prepared to defend it to the exclusion of other points of view. What I am prepared to defend is a broader thesis: there are many special disciplines and several different approaches in biological research; all are important and their integration is most important; to be effective and to be in fact biological all approaches must take into account the *organization* of *organisms,* and so must both depart from and lead to the level of whole organisms.

If one tries to classify the numerous biological disciplines, it quickly becomes apparent that no linear arrangement or single hierarchy will do. The subjects are too diversely characterized and too intricately interrelated in too many different ways for categorization on any one principle. A multidimensional key, with categories of different dimensions or scales intersecting, is required. But I, at least, have not found any of the proposed keys satisfactory. All fail in adequate representation of biology as an integral science and of the interconnections of its constituent disciplines. The value of such attempts is that they do make explicit some categorizations significant of different attitudes and approaches in modern biology.

A dimension that inevitably appears is that of level of organization, a scale running from subatomic particles to multispecific communities. A sharp dichotomy somewhere along that scale is implicit in the now frequent distinction of "molecular biology" and "organismal biology." Another dimension, usually less evident or clear-cut, has to do with the kinds of generalizations or explanations that are sought. That does not involve quite so definite a dichotomy, but it does tend to reinforce the previous dichotomy. There do tend to be differences in approach and aims between infraorganismal and organismal-populational research.

One method of explanation in biology is to relate structure and events at one level of organization with those at lower levels. For example, food chains in a community may, in this sense, be explained successively by the adaptations of specific populations, by the functioning of individual members of each species, by the physiology of their various organ systems, by the enzymes initiating and mediating the underlying chemical and physical processes, by DNA specification of the enzymes, by the structure and properties of DNA molecules, and ultimately by those of the atoms comprising such molecules. That is, of course, the well-known principle of reductionism. In its extreme form, the principle would imply that, for example, the attack of a lioness on a zebra could be not only explained but also predicted from knowledge of the structure and properties of the atoms composing the lioness. Perhaps no one accepts quite so extreme an application, but the first of the statements introducing this chapter certainly approaches it.

In fact, pure reductionism offers no adequate explanations and permits no sound predictions. The action of an atom, a DNA molecule, or an enzyme, can be neither explained nor predicted outside the context of the system in which it occurs. This will not be news to the biophysicists and it is, indeed, involved in my first paraphrase, but biophysicists do often underestimate its force or overlook its significance. It means that explanation by reduction simultaneously implies or must be accompanied by the opposite, which might be called composition: that meaningful biological explanation goes up as well as down the scale of levels of organization. It is just as explanatory and just as essential for understanding to say that lion enzymes digest zebra meat *because* that enables the lion to survive and to perpetuate the population of which he is a part, as to say that the digestive event occurs *because* of specific chemical reactions between lion enzymes and zebra fats, proteins, and so on.

One direction of explanation goes down the scale, and because it leads into their domain it is naturally emphasized by the molecular and other lower-level biologists. The other goes up, and is as naturally emphasized by the organismal and populational biologists. Whatever labels may be adopted for these two approaches or frames of mind, all biological researchers are aware that they exist and that they currently tend to divide many biologists into two often cooperating but sometimes opposed camps.

Reduction and composition are not the only approaches in biology. As in all sciences, research in biology consists of asking a series of questions and seeking answers. Biological questions can be reduced to a few simple and colloquial forms: "What?" "How?" "What for?" "How come?"

"What?" does not elicit an explanation. It suggests the further questions and supplies the data for answering them and checking the answers. It calls for observation, whether in nature or in the laboratory, and it is the question asked in the descriptive branches of biology. Although elementary and (because nonexplanatory) unsatisfying in itself, it is the indispensable basis of all scientific endeavor. It has equal pertinence to all levels of organization and all conceptual approaches.

"How?" means how does this thing work? What makes it go, and how does it function? The functional biologist examines at varying levels such things as the mechanical working of the lion's bones and teeth, the digestive action of his enzymes, the production of those enzymes in cells, and so on. Almost inevitably the direction of research is reductionist, and as that direction is followed the answers tend more and more to be in physical and chemical terms. Ultimately the investigator is not dealing with a lion, or any other organism at all, but with a chemical reaction in a test tube or some form of physical model. This is the typical question of biophysics, biochemistry, molecular biology, cell biology, and the other mainly reductionist, lower-level subsciences of biology. An answer, as has been noted, is one form of explanation, or a partial explanation. This kind of explanation has been called physiological, using that term in a somewhat special way.

"What for?" also relates to function but in a different sense or in the opposite direction. It means, what function or purpose does this serve? What need does it fill in the cell or the individual? What functional effect does it have in the population or the community? In this sense the function of the digestive enzymes is to provide materials and energy for cells (and hence for the whole organism); the function of bones and their muscles is to permit behavior adapted to the needs of the organism; and so on. Almost inevitably the direction of research is compositionist, proceeding from lower to higher levels of organization. Like "How?" the question is appropriate at all levels, but it has an opposite tendency. The answers tend more and more to be in terms of whole organisms, populations, and communities. Ultimately the investigator has no direct concern with the physical sciences but is dealing with phenomena that are exclusively biological. This is a typical question of the biology of whole organisms, or of the compositionist approach at any level. An answer is another form of explanation or another partial explanation. Since there is reference to purpose (fulfilling of needs, functioning in a larger context) which nevertheless is not teleological in the philosophical sense, it has been proposed to call this kind of explanation teleonomic.

"How come?" means how did this originate? What course has it followed through time and what were the causes of that historical development? The question is equally appropriate at all levels, and answers need not inherently be either reductionist or compositionist. We could (if we had enough information) follow the history of DNA without necessary reference to nonmolecular levels, just as we can follow the evolution of species (populations) without necessary reference to nonpopulation levels. Nevertheless, more complete answers tend to cut across all levels and to be simultaneously reductionist, compositionist, and more. The evolution of DNA becomes most fully meaningful and explanatory when it is related to the processes in populations that cause specific information to be coded in the DNA, and, conversely, the evolution of species is best understood when related to the changes in the DNA codes. It is nevertheless true at present that the question "How come?" is more often asked by organismal biologists and that they have produced most of the available answers, even including those involving levels below the individual.

Historical and teleonomic explanations have close conceptual relationships. The most important single element in historical explanation, natural selection, is teleonomic in principle. Confronted by any intricately organized system, anyone inevitably asks both "What for?" and "How come?" and feels that these are mutually dependent problems. For entirely physical systems in nature, the questions "What for?" or "How come?" as distinct from simply "What?" or "How?" often seem purely philosophical or unanswerable and therefore sterile as approaches to scientific research. In biological systems the pervasive fact of evolution makes scientific answers accessible. For these reasons the fields of both teleonomic and historical explanations are often and significantly united under the rubric of evolutionary biology.

The point about explanation in biology that I should particularly like to stress is this: to understand organisms one must explain their organization. It is elementary that one must know what is organized and how it is organized, but that does not explain the fact or the nature of the organization itself. Such explanation requires knowledge of how an organism came to be organized and what functions the organization serves. Ultimate explanation in biology is therefore necessarily evolutionary.

The aim of biology is to understand the structure, functioning (both as "How?" and "What for?"), and history of organisms and of populations of organisms. Biology is the study of life; life occurs only in living things; "living thing" is synonymous with "organism"; organisms that are not whole are not alive. (The expressions "organismal biology" or "biology of whole organisms" are thus really tautologic.) Moreover, organisms live only in the context of populations. Clearly, and even as regards any one phenomenon, biology requires study at all levels of organization and explanations of all three kinds. Nevertheless, the name and nature of the science itself center its aim at the level of organisms in populations. To be truly biological, all approaches, reductionist, compositionist, and historical, must not only include but also depart from or lead to organisms, structured and functioning as such in themselves and in reproducing aggregates.

The whole of this extremely complex goal cannot be achieved or—for the individual biologist, at least—even aimed at all at once. The rate of progress is uneven, and rapid advances take place now in one direction and now in quite another. Once a shove has been given in one direction, perhaps by a technological or conceptual breakthrough, perhaps by individual enthusiasm, perhaps by what seems pure chance, a band wagon effect ensues. Students flock to the accelerating front; money is poured into it; professional advancement, fame, and fortune follow it. That is only natural and is in one respect desirable, for the band wagon effect feeds into a circle by which the rate of discovery in some one field is indeed increased.

Fortunately, more than one aspect of biology may accelerate at the same time, and with the great increase in numbers of biologists not all crowd onto the same band wagon. The gaudiest band wagon just now is manned by reductionists, travels on biochemical and biophysical roads, and carries a banner with a strange device: DNA. There are, however, a good number of other, perhaps less gaudy band wagons, going down other, perhaps less rapid roads, under banners perhaps less vehemently saluted. At any rate, we do salute them all, and yet may fear that these band wagons diminish travel on still other roads, which are falling into neglect but which are also essential to reach the destination toward which all are, or should be traveling.

The danger for biology is the assumption that some of its subjects, some directions of approach, some kinds of explanation are no longer important: their contribution, such as it was, is thought to have been achieved or is not now considered necessary. But the assumption that any of the biological disciplines has fully accomplished its aim or that any is unnecessary to achieve the goal of biology is completely wrong. The error is compounded when it directs attention away from the real center of that goal: the understanding of organisms. All four of the paraphrases introducing this discussion err to the extent that they insist on one approach at the expense of others, but the fourth errs least in that respect, and it has the merit of concentrating attention on the *biologically* most significant level.

It is obviously impossible in this brief chapter to review the particular status of the many separable high-level biological disciplines, even in the most general way. A few examples, necessarily treated without detail, must suffice to suggest the general position of such subjects in current biological research. Classification will serve as one example that is extreme in several respects: apart from strict description, it is the oldest of all biological subjects; it is clearly organismal, although its compositionist approach draws data and explanations from lower organizational levels; and it is now the subject most often considered relatively unimportant by reductionist biologists.

Each biological discipline or subscience has a goal of its own that is part of the total goal of biology. The goal of classification and of taxonomy (the theoretical and operational basis of classification) is to produce a stable, biologically meaningful arrangement of all knowable organisms. To state the goal is to demonstrate that it is not yet anywhere near achievement. Classification is as yet far from stable. Current classifications in some groups have achieved considerable biological significance but could surely be improved. In many other groups, current classifications still have unsound or insignificant foundations. Only a small fraction of knowable organisms is as yet known and classified.

It might seem offhand that classification is essentially a descriptive science, its one question "What?" (What organism is that? An animal; a bird; a passerine; a tyrannid; a phoebe, *Sayornis;* Say's phoebe, *Sayornis saya.*) Even some taxonomists, bemused by confusion of identification and classification, of observation and interpretation, have been that naive. In fact, however, any attempt to arrange the data in a rational or natural way, to classify the observations, brings in some element of explanation. In eighteenth- and early nineteenth-century classifications the explanatory element tended indeed to be minimal or more metaphysical than scientific. Then, in 1859, Darwin made the first great breakthrough by defining or rather demonstrating the application and meaning of "How come?" in this field. Taxonomic orderliness, on which the possibility of meaningful classification depends, is caused by evolution. Darwin started a taxonomic band wagon on its way, enthusiastically, if not too well, manned. By the early twentieth century the impetus was being lost and attention was being diverted to other band wagons, especially one under the banner of Mendel.

In the last generation, through the work of numerous biologists still active, there has been another breakthrough, less abrupt than Darwin's and less sharply definable but with equally important potentialities. Darwin's principle is no less involved, but it is related more especially and explicitly to the fact that it is varying populations that evolve and that are classified. Apparently simple in itself, that principle changes approaches to classification and the actual forms of classification. More importantly, it brings classification into more significant contact with other biological disciplines, and it both intensifies and expands the explanatory content of the subject. "How come?" is still asked, and more complete answers are being found. "How?" becomes equally pertinent. (How do species arise? How are they maintained as separate entities?) "What for?" is hardly less so. (What are the functions of differences among species and other taxa? What is the function of a particular species in its community?) Thus, not only historical but also reductionist and compositionist approaches enter classification, and while continuing to operate at its level of populations it extends upward and downward through all levels. It becomes the core discipline of the science of organic diversity and interrelationships: systematics.

A geneticist friendly to the subject has remarked that taxonomy is currently near the bottom of the biological pecking order. That is perhaps temporarily true in terms of relative acclaim and emphasis. Nevertheless, taxonomy does have a band wagon advancing energetically toward its goal and contributing very significantly to the advance of biology as a whole. The proportionate share of taxonomy in total biological activity just now is below the optimum, but in absolute terms taxonomy is at an all-time peak; in numbers of working taxonomists; in rate of discovery; in accomplishment; in publication; in available support.

As a second example I select biogeography because it is still more obviously at a high organizational level and because it is a complex using the approaches and explanatory methods of several other biological disciplines. Thence it tends to ramify through many fields of biology, and it is among the most highly syndetic of biological subjects. In spite of that fact, it is one of the most poorly supported and most slowly progressing.

In a primitive, prescientific way, biogeography, like classification, dates from prehistoric times, but it belatedly reached fully scientific status only around the middle of the nineteenth century, largely under the influence of Darwin and Wallace. Although Darwin, particularly, aimed at explanatory biogeography and adumbrated almost all later approaches, work in this field was at first largely descriptive. One aim was the production of increasingly generalized descriptions of animal and plant distribution in the form of hierarchies of delimited and denominated biogeographic regions (zoochoria and phytochoria). The higher categories were quickly mapped out, and although there have been many valuable studies of detail in the present century there has been little fundamental advance at that descriptive level.

Explanatory biogeography, most active in the present century and at the present time, has two inseparable but distinguishable aspects. It treats, on one hand, with the requirements (tolerances and preferences) of species of organisms, the geography of environments meeting those requirements, and the consequent distribution of species. The approach is ecological; the questions are mostly "How?" and "What for?"; and explanations are largely reductionist, by individual adaptations within populations, by functioning and functional requirements of organs within individuals, and so on down ultimately to the molecular level. On the other hand, explanatory biogeography also treats of the origins of particular groups of organisms in specified regions and their changes in geographic distribution through time. The approach is historical; the primary question is "How come?"; and explanations are evolutionary.

Geographic considerations enter into many other biological subjects. They are, for instance, crucial in the "How?" of the origin of species (and other taxa), hence also in the "How come?" of systematics. Conversely, although the ecological biogeographer more often asks "How?" he cannot overlook the ulterior "How come?" for the adaptations that explain how.

The status of biogeography is peculiar in that it is rarely pur sued as a discipline or profession in itself. Its data, principles, and methods come mostly from systematics, ecology, and paleontology, and its practitioners are primarily systematists, ecologists, and paleon tologists, not zoogeographers per se. Ecological biogeography has been most intensively developed by botanists, historical biogeography by vertebrate paleontologists. As far as I know there have been no professorships or journals of biogeography in this country, and only one or two anywhere. Development of the subject is de pendent on the personal interests of specialists in related but distinct disciplines, and support for it is largely incidental or marginal. Un der one aegis or another, it is making progress, but that progress is slower and more erratic than would be desirable. No band wagon flies this banner.

As a final, brief example I select a subject that is just as much infraorganismal as organismal and populational, that currently has some tendency to split along those lines, and that by the irrationality of that fission demonstrates the artificiality of the dichotomous tendency in biology and the impracticality of emphasizing one approach at the expense of others.

Modern genetics started at the level of individual organisms, individuals taken collectively and studied statistically (that was of course Mendel's great contribution to methodology in the study of heredity) but not considered as biological populations. Although many geneticists now use the term to include all study of biparental heredity, Mendelian genetics, strictly speaking and in the original sense, is concerned with the inheritance of particular characters in individuals. Explanation is physiological, and approach is largely reductionist, e.g., from organisms to chromosomes to genes, the latter (as originally defined) deduced but not observed. This field is still extremely active, although no longer so clearly delimited in subject matter or in approach.

Population genetics focused attention on the next higher organizational level, and it complemented the original Mendelian approach by balancing reduction with composition, physiological with teleonomic explanation. The teleonomic aspect is involved in the explanation of genetic information in organisms by adaptive processes, notably those of differential reproduction, or natural selection in the modern sense, at the population level. Thus genetics was able to reach more complete explanation by asking not only "How?" but also "What for?" Systematics was as instrumental as genetics in reaching this more fully explanatory combined approach, and the present synthetic theory of evolution was born from the union of the two disciplines. In conjunction with paleontology, "How come?" is also asked and, in part, answered. (The theory turns out to be tri-parental, at least!)

Population genetics was unusual, if not unique, in biology in that it was largely developed by deduction from mathematical models. Its recent tendency has been to outgrow the models, or to become too complex for that method alone, and to proceed more extensively by induction from observation in field and laboratory. The development of the synthetic theory was a major breakthrough, and population genetics in that framework and in company with systematics, paleontology, and other related disciplines continues to be extremely active and exuberantly productive.

Most recently, the original tendency of Mendelian genetics has carried another line of genetical research more and more into the realm of reductionist biology. There, a breakthrough in molecular biology had occurred, and there has been a meeting ground and a deepening of the reductionist part of genetical explanation. This centers on DNA and emphasizes genetic processes within cells and at the molecular level, including nonchromosomal heredity. The extent of enthusiastic emphasis in some quarters is suggested by my third introductory paraphrase, which implies, at least, that genetics as a whole can be reduced to chemistry and that organisms can be explained by chemical genetics, without in turn helping to explain the latter and without explanations also from and to other levels.

I do not at this point need to restate all the reasons why I disagree with that point of view and consider it a hindrance in current advance toward the proper goal of biology. Explanation in genetics and in the whole field of the biological sciences must answer all the questions, use all approaches, and recognize that all biological disciplines are inextricably interdependent. Organismal and population genetics can no more reach their goal without consideration of DNA than molecular and cell genetics can reach theirs without taking whole organisms and populations into account.

Systematics, biogeography, and population genetics are far from being the only biological disciplines involving organismal and higher levels of organization, largely compositionist in approach, and characterized by teleonomic and historical explanations. Among other subjects primarily at these levels are paleontology, ecology, and behavior. Still more subjects have essential aspects that are organismal in the same sense, notably comparative anatomy, physiology, and embryology. All those disciplines are no less, and I believe no more, important than the three I have used as examples. It would be contrary to my whole thesis to suggest that these organismal or evolutionary aspects of biology should be stressed at the expense of any others or pursued without reference to others. It is, however, my conviction that this general group of disciplines is focal for biology and that it now offers the greatest opportunities for integration and progress in fully *biological* science.

CHAPTER SEVEN

The Historical Factor in Science

THE special problems of evolutionary biology involve a historical factor not, indeed, unique to biology, but highly characteristic of it. Historical explanation of the "How come?" sort does occur quite widely among the sciences and raises questions that go beyond the classical philosophical concepts of reductionist and physical science. Are there historical laws? Can scientific testing be nonpredictive? Can unique events be explained, or in any way considered scientifically? What is history, anyway, and is there historical science? Those and related questions, some of them implicit in the last two chapters, are now to be more closely examined. The first concern will be to define historical science, which depends in turn on definitions of history and of science.

The simplest definition of history is that it is change through time. It is, however, at once clear that the definition fails to make distinctions necessary if history is to be studied in a meaningful way. A chemical reaction involves change through time, but obviously it is not historical in the same sense as the first performance by Lavoisier of a certain chemical experiment. The latter was a nonrecurrent event, dependent on or caused by antecedent events in the life of Lavoisier and the lives of his predecessors, and itself causal of later activities by Lavoisier and his successors. The chemical reaction involved has no such causal relationship and has undergone no change before or after Lavoisier's experiment. It always has occurred and always will recur under the appropriate historical circumstances, but as a reaction in itself it has no history.

A similar contrast between the historical and the nonhistorical exists in biology and other sciences. The chemical reactions and physical processes in cells are indefinitely repeatable, unchanging in character, and nonhistorical. Each real, individual organism at a given time is unique and changes through time to other unique, non-recurrent configurations. Those individual configurations are historical, while the physical and chemical properties and processes by which they change are not.

The unchanging properties of matter and energy and the likewise unchanging processes and principles arising therefrom are *immanent* in the material universe. They are nonhistorical, even though they occur and act in the course of history. The actual state of the universe or of any part of it at a given time, its configuration, is not immanent and is constantly changing. It is *contingent,* in Bernal's (1951) term, or *configurational,* as I prefer to say (Simpson, 1960). History may be defined as configurational change through time, a sequence of real, individual but interrelated events. These distinctions between the immanent and the configurational and between the nonhistorical and the historical are essential to clear analysis and comprehension of history and of science. They will be maintained, amplified, and exemplified in what follows.

In Chapter 5 I defined science as an exploration of the material universe that seeks natural, orderly relationships among observed phenomena and that is self-testing. Apart from the points that science is concerned only with the material or natural and that it rests on observation, the definition involves three scientific activities: the description of phenomena, the seeking of theoretical, explanatory relationships among them, and some means of establishing confidence regarding observations and theories. Historical science, then, may be defined as the determination of configurational sequences, their explanation, and the testing of such sequences and explanations.

So long as biology is concerned with the identification and description of chemical processes or of the state of the present world of living things, it is essentially nonhistorical. However, if it remains at those mainly descriptive levels it also falls short of the full definition of science. As soon as fully explanatory relationships are brought in, so must be the changes and sequences of configurations, that is, the historical factor. That factor thus is involved in the thorough study of all aspects of biology, and not only of those obviously and directly historical in form.

The biological subject most obviously and directly historical is paleontology, which is also a part of the subject matter of historical geology. In its biological role, like all other aspects of biology, it involves all the immanent properties and processes of the physical sciences, but differs from them not only in being historical but also in that its configurational systems are incomparably more complex and have feedback and information storage and transmittal mechanisms unlike any found in the inorganic realm. Its involvement in geology and inclusion in that science as well as in biology is primarily due to the fact that the history of organisms runs parallel with, is environmentally contained in, and continuously interacts with the physical history of the earth. It is of less philosophical interest but of major operational importance that paleontology, when applicable, has the highest resolving power of any method yet discovered for determining the sequence of strictly geological events. (That radiometric methods may give equal or greater resolution is at present a hope and not a fact.)

In principle, the observational basis of any science is a straight description of what is there and what occurs, what Lloyd Morgan used to call "plain story." In a physical example, plain story might be the specifications of a pendulum and observations of its period. Geological plain story might describe a bed of arkose, its thickness, its attitude, and its stratigraphic and geographic position. An example of paleontological plain story would be the occurrence of a specimen of a certain species at a particular point in the bed of arkose. In general, the more extended plain stories of historical science would describe configurations and place them in time.

In fact, plain story in the strictest, most literal sense plays little part in science. Some degree of abstraction, generalization, and theorization usually enters in even at the first observational level. The physicist has already abstracted a class of configurational systems called "pendulums" and assumes that only the length and period need be observed, regardless of other differences in individuals of the class, unless an observation happens to disagree with the assumption. Similarly, the geologist by no means describes all the characteristics of the individual bed of arkose and its parts but has already generalized a class "arkose" and adds other details, if any, only in terms of such variations within the class as are considered pertinent to his always limited purpose. The paleontologist has departed still further from true, strict plain story, for in recording a specimen as of a certain species he has not only generalized a particularly complex kind of class but also reached a conclusion as to membership in that class that is not a matter of direct observation at all.

Every object and every event is unique if its configurational aspects are described in full. Yet, and despite the school teachers, it may be said that some things are more unique than others. This depends in the first place on the complexity of what is being described, for certainly the more complex it is the more ways in which it may differ from others of its general class. A bed of arkose is more complex than a pendulum, and an organism is to still greater degree more complex than a bed of arkose. That hierarchy of complexity and individual uniqueness from physics to geology to biology is characteristic of those sciences and essential to philosophical understanding of them. It bears on the degree and kind of generalization characteristic and appropriate to the various sciences even at the primary observational level. The number of pertinent classes of observations distinguished in physics is much smaller than in geology, and much smaller in geology than in biology. For instance, in terms of taxonomically distinguished discrete objects, compare the numbers of species of particles and atoms in physics, of minerals and rocks in geology, and of organisms in biology. Systems and processes in these sciences have the same sequence as to number and complexity.

Another aspect of generalization and degree of uniqueness arises in comparison of nonhistorical and historical science and in the contrast of immanence and configuration. In the previous examples, the physicist was concerned with a nonhistorical and immanent phenomenon: gravitation. It was necessary to his purpose and inherent in his method to eliminate as far as possible and then to ignore any historical element and any configurational uniqueness in the particular, individual pendulum used in the experiment. He sought a changeless law that would apply to all pendulums and ultimately to all matter, regardless of time and place. The geologist and paleontologist were also interested in generalization of common properties and relationships, between one occurrence of arkose and another, between one specimen and another of a fossil species, but their generalizations were of the configurational and not the immanent and were, or at least involved, historical and not only nonhistorical science. The arkose or the fossil had its particular as well as its general configurational properties, its significant balance of difference and resemblance, not only because of immanent properties of its constituents and immanent processes that had acted on it, but also because of its history, the configurational sequence by which these individual things arose. The latter aspect, not pertinent to the old pendulum experiment or to almost anything in the more sophisticated physics of the present day, is what primarily concerns geology and paleontology as historical sciences, or historical science in general.

It has been mentioned that the purpose of the pendulum experiment was to formulate a law. The concept of scientific law and its relationship with historical and nonhistorical science is a disputed question requiring clarification. The term "law" has been so variously and loosely used in science that it is no longer clear unless given an explicit and restrictive definition. The dictionary that I happen to have at hand defines law in philosophical or scientific use as "a statement of a relation or sequence of phenomena invariable under the same conditions." That is satisfactory if it is made clear that a law applies to phenomena that are themselves variable: it is the *relationship* (or sequence, also a relationship) that is invariable. "Under the same conditions" must be taken to mean that other variables, if present, are in addition to and not inextricably involved with those specified in the law. It is further perhaps implicit but should be explicit that the relationship must be manifested or repeatable in an indefinitely recurrent way. A relationship that could or did occur only once would indeed be invariable, but surely would not be a law in any meaningful scientific sense. With those considerations, the definition might be rephrased thus: a scientific law is a recurrent, repeatable relationship between variables that is itself invariable to the extent that the factors affecting the relationship are explicit in the law.

The definition implies that a valid law includes all the factors that *necessarily* act in conjunction. The fact that air friction also significantly affects the acceleration of a body falling in the atmosphere does not invalidate the law of gravitational acceleration, but only shows that the body is separately acted on by another law. Friction and the factors of gravitational acceleration are independent. Both laws are valid, and they can be combined into a valid compound law. But if some factor *necessarily* involved in either one, such as force of gravitation for acceleration or area for friction, were omitted, that law would be invalidated.

Laws, as thus defined, are generalizations, but they are generalizations of a very special kind. They are complete abstractions from the individual case. They are not even concerned with what individual cases have in common, in the form of descriptive generalizations or definitions such as that all pendulums are bodies movably suspended from a fixed point, all arkoses are sedimentary rocks containing feldspar, or all vertebrates are animals with jointed backbones. Those and similar generalizations are obviously not laws by any usage. When we say, for instance, that arkose is a feldspathic sedimentary rock, we mean merely that we have agreed that if a rock happens to be sedimentary and to be within a certain range of texture and of composition including a feldspar, we will call it "arkose." We do not mean that the nature of the universe is such that there is an inherent relationship among sedimentary rocks and feldspars reducible to a constant. Laws, on the contrary, are inherent, that is *immanent,* in the nature of things as abstracted entirely from contingent configurations, although always acting on those configurations.

Until recently the theoretical structure of the nonhistorical physical sciences consisted largely of a body of laws or supposed laws of this kind. The prestige of those sciences and their success in discovering such laws were such that it was commonly believed that the proper scientific goal of the historical sciences was also to discover laws. Supposed laws were proposed in all the historical sciences. By way of example in paleontology, I may mention "Dollo's law" that evolution is irreversible, "Cope's law" that animals become larger in the course of evolution, or "Williston's law" that repetitive serial structures in animals evolve so as to become less numerous but more differentiated. Most such supposed laws are no more than descriptive generalizations. For example, animals do not invariably become larger in time. "Cope's law" merely generalizes the observation that this is a frequent tendency, without establishing any fixed relationship among variables possibly involved.

Even when a relationship seems established, so called historical "laws" are almost always open to exceptions. For example, Rensch, an evolutionist convinced of the validity of historical laws, considers "Allen's rule" a law that when mammals adapt to colder climates their feet become shorter. But of the actual mammals studied by him, 36 per cent were exceptions to the "law." Rensch explains this by supposing that "many special laws act together or interfere with one another. Thus 'exceptions' to the laws result." This is a hypothetical possibility, but to rely upon it is an act of faith. The "interfering" laws are unknown in this or similar examples. A second possibility is that the "laws," as stated, are invalid as laws because they have omitted factors necessarily and inherently involved. I believe that this is true, not in the sense that we have only to complete the analysis and derive a complete and valid law, but in the sense that the omissions are such as to invalidate the very concept of historical law.

The search for historical laws is, I maintain, mistaken in principle. Laws apply, in the dictionary definition "under the same conditions," or in my amendment "to the extent that factors affecting the relationship are explicit in the law," or in common parlance "other things being equal." But in history, a sequence of real, individual events, other things never are equal. Historical events, whether in the history of the earth, the history of life, or recorded human history, are determined by the immanent characteristics of the universe acting on and within particular configurations, and never by either the immanent or the configurational alone. It is a law that states the relationship between the length of a pendulum—*any* pendulum—and its period. Such a law does not include the contingent circumstances, the configuration, necessary for the occurrence of a real event, say Galileo's observing the period of a particular pendulum. If laws thus exclude factors inextricably and significantly involved in real events, they cannot belong to historical science.

It is further true that historical events are unique, usually to a high degree, and hence cannot embody laws defined as recurrent, repeatable relationships. Apparent repetition of simple events may seem to belie this. A certain person's repeatedly picking up and dropping a certain stone may seem to be a recurrent event in all essentials, but there really is no applicable *historical* law. Abstraction of a law from such repeated events leads to a nonhistorical law of immanent relationships, perhaps in this case of gravity and acceleration or perhaps of neurophysiology, and not to a historical law of which this particular person, picking up a certain stone, at a stated moment, and dropping it a definite number of times would be a determinate instance. In less trivial and more complex events, it is evident that the extremely intricate configurations involved in and necessary, for example, as antecedents for the erosion of the Grand Canyon or the origin of *Homo sapiens* simply cannot recur that there can be no laws of such one-of-a-kind events. (Please bear in mind that the true, immanent laws are equally necessary and involved in such events but that they remain nonhistorical; the laws would have acted differently and the historical event, the change of configuration, would have been different if the configuration had been different; this historical element is not included in the operative laws.)

It might be maintained that my definition of law is old-fashioned and is no longer accepted in the nonhistorical sciences either. Many laws of physics, considered nonhistorical, are now conceived of as statistical in nature, involving not an invariable relationship but an average one. The old gas laws or the new laws of radioactive decay are examples. The gas laws used to assume an ideal gas. Now they are recognized as assuming that directions of molecular motion tend to cancel out if added together and that velocities tend to vary about a mean under given conditions. That cannot be precisely true of a real gas at a given moment, but when very large numbers of molecules are involved over an appreciable period of time the statistical result is so close to the assumptions that the gas laws hold as near as does not matter. Thus and in similar ways the descent from the ideal to the real in physical science has been coped with, not so much by facing it as by finding devices for ignoring it.

The historical scientist here notes that a real gas in a real experiment has *historical* attributes that are *additional* to the laws affecting it. Every molecule of a real gas has its individual history. Its position, direction of motion, and velocity at a given moment (all parts of the total configuration) are the outcome of that history. It is, however, quite impractical and for the purposes of physics unnecessary to make a historical study of the gas. The gas laws apply well enough "other things being equal," which means here that the simple histories of the molecules tend, as observation shows, to produce a statistical result so nearly uniform that the historical, lawless element can be ignored for practical purposes.

The laws immanent in the material universe are not statistical in essence. They act invariably in variable historical circumstances. The pertinence of statistics to such laws as those of gases is that they provide a generalized description of usual historical circumstances in which those laws act, and not that they are inherent in the laws themselves. Use of statistical expressions not as laws but as generalized descriptions is common and helpful in all science and especially in historical science. For example, the statistical specifications of variation in populations of organisms clearly are not laws but descriptions of configurations involved in and arising from history.

To speak of "laws of history" is either to misunderstand the nature of history or to use "laws" in an inacceptable sense, usually for generalized descriptions rather than formulations of immanent relationships.

Uniformitarianism has long been considered a basic principle and sometimes claimed to be a law of historical science and a major contribution of geology to science and philosophy. In one form or another it does permeate historical thought to such a point as often to be taken for granted. Among those who have recently given conscious attention to it, great confusion has arisen from conflicts and obscurities as to just what the concept is. To some, uniformitarianism (variously defined) is a law of history. Others, maintaining that it is not a law, have tended to deny its significance. Indeed, in any reasonable or usual formulation, it is not a law, but that does not deprive it of importance. It is commonly defined as the principle that the present is the key to the past. That definition is, however, so loose as to be virtually meaningless in application. A new, sharper, and clearer definition in modern terms is needed.

Uniformitarianism arose around the end of the eighteenth century, and its original significance can be understood only in that context. (The historical background is well covered in Gillispie, 1951.) It was a reaction against the then prevailing school of catastrophism, which had two main tenets: (1) the general belief that God has intervened in history, which therefore has included both natural and supernatural (miraculous) events; and (2) the particular proposition that earth history consists in the main of a sequence of major catastrophes, usually considered as of divine origin in accordance with the first tenet. Uniformitarianism, as then expressed, had various different aspects and did not always face these issues separately and clearly. On the whole, however, it embodied two propositions contradictory of catastrophism: (1) earth history (if not history in general) can be explained in terms of natural forces still observable as acting today; and (2) earth history has not been a series of universal or quasi-universal catastrophes but has in the main been a long, gradual development—what we would now call an evolution. (The term "evolution" was not then customarily used in this sense.) A classic example of the conflicting application of these principles is the catastrophist belief that valleys are clefts suddenly opened by a supernally ordered revolution as against the uniformitarian belief that they have been gradually formed by rivers that are still eroding the valley bottoms.

Both of the major points originally at issue are still being argued on the fringes of science or outside it altogether. To most geologists, however, they no longer merit attention from anyone but a student of human history. It is a necessary condition and indeed part of the definition of science in the modern sense that only natural explanations of material phenomena are to be sought or can be considered scientifically tenable. It is interesting and significant that general acceptance of this principle (or limitation, if you like) came much later in the historical than in the nonhistorical sciences. In historical geology it was the most important outcome of the uniformitarian-catastrophist controversy. In historical biology it was the still later outcome of the Darwinian controversy and was hardly fully settled until our own day. (It is still far from settled among nonscientists.)

As to the second major point originally involved in uniformitarianism, there is no a priori or philosophical reason for ruling out a series of *natural* worldwide catastrophes as dominating earth history. That is simply in such flat disagreement with everything we now know of geological history as to be completely incredible. The only issues still valid involve the way in which natural processes still observable have acted in the past and the sense in which the present is a key to the past. Uniformitarianism, or neo-uniformitarianism, as applied to those issues has taken many forms, among them two extremes that are both demonstrably invalid. They happen to be rather amusingly illustrated in a recently published exchange of letters by Lippman and Farrand.

Lippman attacks uniformitarianism on the assumption that its now "orthodox" form is absolute gradualism: belief that geological processes have *always* acted gradually and that changes catastrophic in rate and extent have never occurred. Farrand demonstrates that Lippman has set up a straw man. Catastrophes do now occur. Their occurrence in the past exemplifies rather than contradicts a principle of uniformity. It happens that there is no valid evidence that catastrophes of the kind and extent claimed by the original catastrophists and by Lippman have ever occurred or that they could provide explanations for some real phenomena as claimed. That, however, is a different point. Farrand expresses a common, probably the usual, modern understanding of uniformitarianism as "the geologist's concept that processes that acted on the earth in the past are the *same processes* that are operating today, on the *same scale* and at approximately the *same rates*" (italics mine). But that principle is also flatly contradicted by geological history. Some processes (those of vulcanism or glaciation, for example) have evidently acted in the past with scales and rates that cannot by any stretch be called "the same" or even "approximately the same" as those of today. Some past processes (such as those of Alpine nappe formation) are apparently not acting today, at least not in the form in which they did act. There are innumerable exceptions that disprove the rule.

Then what uniformity principle, if any, is valid and important? The distinction between immanence and configuration (or contingency) clearly points to one: the postulate that immanent characteristics of the material universe have not changed in the course of time. By that postulate all the immanent characteristics exist today and so can, in principle, be observed or, more precisely, inferred as generalizations and laws from observations. It is in this sense that the present is the key to the past. Present immanent properties and relationships permit the interpretation and explanation of history precisely because they are *not* historical. They have remained unchanged, and it is the configurations that have changed. Past configurations were never quite the same as they are now and were often quite different. Immanent characteristics working within those different configurations therefore produced changes that necessarily had at various times scales, rates, and combinations into complex processes different from those of today. The uniformity of the immanent helps to explain the fact that history is not uniform. Only to the extent that past configurations resembled the present in essential features can past processes have worked in a similar way.

That immanent characteristics are unchanging may seem at first sight either a matter of definition or an obvious conclusion, but it is neither. Gravity would be immanent (an inherent characteristic of matter *now*) even if the law of gravity had changed, and it is impossible to prove that it has not changed. Uniformity, in this sense, is an unprovable postulate justified, or indeed required, on two grounds. First, nothing in our incomplete but extensive knowledge of history disagrees with it. Second, only on this postulate is a rational interpretation of history possible, and we are justified in seeking—as scientists we *must* seek—such a rational interpretation. It is on this basis that I have assumed on previous pages that the immanent is unchanging.

Although the issue of uniformitarianism first arose in geology and has usually been discussed in that context, it is even more crucial for study of the history of life. Only on this basis can we hope for historical explanations of biological phenomena.

Explanation is an answer to the question "Why?" But as Nagel has shown at length, this is an ambiguous question calling for fundamentally different *kinds* of answers in various contexts. One kind of answer specifies the inherent necessity of a proposition, and those are the answers embodied in laws. Some philosophers insist that this is the only legitimate form of explanation. Some, such as Hobson, even go so far as to maintain that since inherent necessity cannot be *proved* there is no such thing as scientific explanation. Nagel demonstrates that all this is in part a mere question of linguistic usage and to that extent neither important nor interesting. The only substantial question involved is whether explanation must be universal or may be contingent. Nagel further shows, with examples, that contingent explanations are valid in any usual and proper sense of the word "explanation." Nagel does not put the matter in just this way and he makes other distinctions not pertinent just here, but in essence this distinction of universal and contingent explanation parallels that between, on one hand, immanence and nonhistorical science, which involves laws, and, on the other, configuration and historical science, which does not involve laws but which does also have explanations.

The question "Why?" can be broken down into three others, each evoking a different kind of explanation, as has been exemplified in previous chapters. "How?" is the typical question of the nonhistorical sciences. It asks how things work: how streams erode valleys, how mountains are formed, how animals digest food—all in terms of the physical and chemical processes involved. The first step toward explanation of this kind is usually a generalized description, but answers that can be considered complete within this category are ultimately in the form of laws embodying invariable relationships among variables. It is at this level that nonhistorical scientists not only start but usually also stop.

The historical scientists nevertheless go on to a second kind of explanation that is equally scientific and ask a second question, in the vernacular "How come?" How does it happen that the Colorado River formed the Grand Canyon or that lions live on zebras? Again the usual approach is descriptive, the plain-story history of changes in configurations, whether individual, as for the Grand Canyon, or generalized to some degree, as for the concurrent evolution of lions and zebras. This is already a form of explanation, but full explanation at this more complex level is reached only by combination of the configurational changes with the immanent properties and processes present within them and involved in those changes. One does not adequately explain the Grand Canyon either by describing the structure of that area and its changes during the Cenozoic or by enumerating the physical and chemical laws involved in erosion, but by a combination of the two.

Two other kinds of scientific explanation are more strictly biological and psychological. Both are kinds of answers to the question "What for?" This question is inappropriate in the physical sciences or the physical ("How?") aspects of other sciences, historical or nonhistorical. "What does a stone fall for?" or "Why was the

Grand Canyon formed?" (in the sense of "What is it now for?") are questions that make no sense to a modern scientist. Such questions were nevertheless asked by primitive scientists (notably Aristotle) and are still asked by some nonscientists and pseudoscientists. The rise of modern physical science required the rejection of this form of explanation, and physical scientists insisted that such questions simply *must not* be asked. In their own sphere they were right, but the questions are legitimate and necessary in the life sciences.

One kind of "What for?" question calls for a teleonomic answer: "What are birds' wings for?" That they are an adaptation to flying is a proper answer and partial explanation near the descriptive level. Fuller explanation is historical: through a sequence of configurations of animals and their environments wings became possible, had an advantageous function, and so evolved through natural selection. Such a history is possible only in systems with the elaborate feedback and information-storage mechanisms characteristic of organisms, and this kind of explanation is inapplicable to wholly inorganic systems (or other configurations). "What for?" may also be answered teleologically in terms of purpose, explaining a sequence of events as means to reach a goal. Despite Aristotle and the Neo-Thomists, this form of explanation is scientifically legitimate only if the goal is foreseen. It therefore is applicable only to the behavior of humans and, with increasing uncertainty, some other animals.

The question "How come?" is peculiar to historical science and necessary in all its aspects. Answers to this question are *the* historical explanations. Nevertheless the full explanation of history requires *also* the reductionist explanations (nonhistorical in themselves) elicited by "How?" Teleonomic explanations are also peculiar to historical science, but only to that part of it dealing with the history of organisms.

Let us now consider testing and prediction in relationship to the historical factor in science. All of science rests on postulates that are not provable in the strictest sense. The uniformity of the immanent, previously discussed, is only one such postulate, although perhaps the most important one for historical science. Indeed it may be said that not only the postulates but also the conclusions of science, including its laws and other theories, are not strictly provable. Proof in an absolute sense occurs only in mathematics or logic when a conclusion is demonstrated to be tautologically contained in axioms or premises. Since those disciplines are not directly concerned with the truth or probability of axioms or premises, and hence of conclusions drawn from them, their proofs are trivial for the philosophy of the natural sciences. In these sciences, the essential point is determination of the probability of the premises themselves, and mathematics and logic only provide methods for correctly arriving at the implications contained in those premises. Despite the vulgar conception of "proving a theory," which does sometimes creep into the scientific literature, careful usage never speaks of proof in this connection but only of establishment of degrees of confidence.

In the nonhistorical sciences the testing of a proposition, that is, the attempt to modify the degree of confidence in it, usually has one general form. A possible relationship between phenomena is formulated on the basis of prior observations. With that formulation as a premise, implications as to phenomena not yet observed are arrived at by logical deduction. In other words, a prediction is made from a hypothesis. An experiment is then devised in order to determine whether the predicted phenomena do in fact occur. The premise as to relationships, the hypothesis, often has characteristics of a law, although it may be expressed in other terms. As confidence increases (nothing contrary to prediction is observed) it becomes a theory, which is taken as simultaneously explaining past phenomena and predicting future ones.

Physical scientists (e.g., Conant) have often maintained or assumed that this is the paradigm of testing ("verification" or increase of confidence) for science in general. On that basis, some philosophers and logicians of science (notably Hempel and Oppenheim) have concluded that scientific explanation and prediction are inseparable. Explanation (in this sense) is a correlation of past and present; prediction is a correlation of present and future. The tense does not matter, and it is maintained that the logical characteristics of the two are the same. They are merely two statements of the same relationship. That conclusion is probably valid as applied to scientific laws, strictly defined, in nonhistorical aspects of science. In previous terms, it has broad—perhaps not completely general— validity for "How?" explanations. But we have seen that there are other kinds of scientific explanations and that some of them are more directly pertinent to historical science. It cannot be assumed and indeed will be found untrue that parity of explanation and prediction is valid in historical science.

Scriven, in published works and personal communication, has discussed this matter at length. One of his points (put in different words) is that explanation and prediction are not necessarily symmetrical, that in some instances a parity principle is clearly inapplicable to them. Part of the argument may be paraphrased as follows. If X is always preceded by A, A is a cause, hence at least a partial explanation, of X. But A may not always be followed by X. Therefore although A explains X when X does occur, it is not possible to predict the occurrence of X from that of A. A simple geological example (not from Scriven) is that erosion causes valleys, but one cannot predict from the occurrence of erosion that a valley will be formed. In fact quite the contrary may occur: erosion can also obliterate valleys.

The example also illustrates another point by Scriven (again in different terms). The failure of prediction is due to the fact that erosion (A) is only a partial cause of valleys (X). It is a (complex) immanent cause, and we have omitted the configurational cause. Erosion is always followed by a valley formation, A is followed by X, if it affects certain configurations. The total cause, as in all historical events, comprises both immanent and configurational elements. It further appears that prediction is possible in historical science, but only to a limited extent and under certain conditions. If the immanent causation is known and if the necessary similarities of configurational circumstances are known and are recurrent, prediction is possible.

The possibility of predicting the future from the past is nevertheless extremely limited in practice and incomplete even in principle. There seem to be four main reasons for these limitations. Mayr has discussed them in connection with historical aspects of biology.

1. A necessary but insufficient cause may not be positively correlated with the usual outcome or event. This is related to the asymmetry of explanation and prediction already discussed, and it is also discussed in other words by Scriven. Scriven's example is that paresis is caused by syphilis, but that most syphilitics do not develop paresis. A modification of Mayr's example is that mutation is necessary for evolutionary change, but that such change rarely takes the direction of the most frequent mutations.

2. The philosophical interest of the foregoing reason for historical unpredictability is reduced by the fact that the outcome might become predictable in principle if *all* the necessary causes were known. But as soon as we bring in configuration as one of the necessary causes, which must always be done in historical science, the situation may become extremely, often quite impossibly complicated. Prediction is possible only to the extent that correlation can be established with pertinent, abstracted and generalized, recurrent elements in configurations. In our geological example, considerations as to base level, slope, precipitation, and other configurational features may be generalized so as to permit prediction that *a* valley will be formed. It would be impossibly difficult to specify all the far more complex factors of configuration required to predict the exact form of a particular valley, an actual historical event. In such cases it may still be possible, as Scriven has pointed out in a different context, to *recognize* a posteriori the configurational details responsible for particular characteristics of the actual valley, even though those characteristics were not practically predictable. This reason for unpredictability of course becomes more important the more complex the system involved. As both Scriven and Mayr emphasize, it may become practically insurmountable in the extremely complex organic systems involved in evolution, and yet this does not make evolution inexplicable. Even in the comparatively extremely simple physical example of the gas laws, it is obviously impossible in practice and probably also in principle (because of the limitations of simultaneous observation of position and motion) to determine the historical configurations of all the individual molecules, so that the *precise* outcome of a *particular* experiment is in fact unpredictable.

In that example the complications may be virtually eliminated and in historical science they may often be at least alleviated by putting specification of configurational causes on a statistical basis. That may, however, still further increase the asymmetry of explanation and prediction. For instance in Scriven's previously cited example, as he points out, the only valid *statistical* prediction is that syphilis will not produce paresis, in other words that a necessary cause of a particular result will *not* have that result. If, as a historical fact, a syphilitic does become paretic, the event was not predictable even in principle. The point is pertinent here as demonstrating that a statistical approach does not eliminate the effect of configurational complication in making historical events unpredictable.

3. As configurational systems become more complex they acquire characteristics absent in the simpler components of those systems and not evidently predictable from the latter. This is the often discussed phenomenon of emergence. The classical physical example is that the properties of water may be explicable but are not predictable by those of hydrogen and oxygen. Again the unpredictability increases with configurational complications. It is difficult to conceive prediction from its component atoms to a mountain range, and to me, at least, prediction from atoms to, say, the fall of Rome, is completely inconceivable. It could be claimed that prediction of emergent phenomena would be possible if we really knew *all* about the atoms. This might just possibly, and only in principle, be true in nonhistorical science, as in the example of $2H + O \rightarrow H_2O$. It would, however, be true in historical science only if we knew all the immanent properties and *also* all the configurational histories of all the atoms, which is certainly impossible in practice and probably in principle. Whether or not the predictability of emergent phenomena is a philosophical possibility (and I am inclined to think it is not), that possibility would seem to have little heuristic and no pragmatic value.

4. Scientific prediction depends on recurrence or repeatability. Precise prediction of unique events is impossible either in practice or in principle. Historical events are always unique in some degree, and they are therefore never precisely predictable. However, as pre-

viously noted, there are different degrees of uniqueness. Historical events may therefore be considered predictable in principle to the extent that their causes are known and are similar. (This is a significant limitation only for configurational causes, since on the postulate of uniformity the immanent causes are not merely similar but identical.)

In practice further severe limitations are imposed by the difficulties of determining what similarities of cause are pertinent to the events and of observing those causal factors. It must also again be emphasized that such prediction can only be general and not particular. In other words, it does not include any unique aspect of the event, and in historical science it is often the unique aspects that most require explanation.

That also raises the point of what is interesting or significant in a scientific investigation. In the physical study of gases or of sand grains, the individuality (uniqueness) of single molecules or grains, slight in any case, is generally beside the point. In dealing with historical events, and especially when these concern organisms, individuality often is just the point at issue. Here, more or less parenthetically, another aspect and use of the statistical approach is pertinent. A statistical description of variation in members of a species is a practical means of taking into account their individual contributions to the over-all individuality of the species.

The most common form taken by attempts at actual historical prediction is the extrapolation of trends. In fact this has no philosophical and little pragmatic validity. Its philosophical justification would require that contingent causes be unchanging or change always in the same ways, which observation shows to be certainly false. Its degree of pragmatic justification depends on the fact that trends and cycles do exist and (by definition) continue over considerable periods of time. Therefore at randomly distributed times established trends and cycles are more likely to continue than not. However, that is true only for short-range prediction, and the likelihood decreases for longer ranges until the appropriate statistical prediction becomes not continuation but termination or change of trend or cycle within the specified time.

The period of likely continuation or justifiable extrapolation is, furthermore, greatly reduced by the fact that a trend or cycle must *already* have gone on for a considerable time in order to be recognized as actually existent. Present knowledge of evolutionary history suggests that all known trends and cycles have in fact ended or changed except those now still within the span of likelihood statistically indicated by those of the past. Moreover, many supposed examples, such as the supposed trend for increase in size of machairodont sabers, now seem to have been mistaken. Many real trends and cycles also turn out to be neither so uniform nor so long continued as was formerly supposed, often under the influence of invalid historical "laws" such as that of orthogenesis. It is questionable whether prediction about a total historical situation on the basis of trends alone is ever justified, even when prediction from causal properties and configurations is possible within limits.

The sequence hypothesis-prediction-experiment is not the only strategy of explanation and testing in nonhistorical physical science. It is, however, so often appropriate and useful there that philosophers who base their concepts of science on physical science, as most of them do, tend to consider it ideal if not obligatory. (On this point of view see, as a single example among many, *Scientific Explanation*, by R. B. Braithwaite.) This is an example of the existing hegemony of the physical sciences, which is not logically justifiable but has been fostered by human historical and pragmatic factors. It has been shown that this strategy is also possible in historical science, but that it here plays a smaller and less exclusive role. It must be supplemented and frequently supplanted by other strategies. These are in part implicit in what has already been said, but further notice of some of the more important of them remains as the final aim of this chapter. One purpose is to demonstrate more fully that nonpredictive explanation and testing are in fact possible in historical geology and other historical sciences.

The primary data of the historical scientist consist of partial descriptions of configurations near the level of plain story. If the configurations are sequential and connected, that is if the later historically arose from the earlier, the antecedent can be taken as including, at least in part, the configurational requirements and causes for the consequent. Even in such simple circumstances, a direct causal connection can often be assumed on the basis of principles already developed or of known parallels. For instance partial configurational causation is clearly involved in the sequence *Hyracotherium* (eohippus)–*Equus* or sand–sandstone. The latter example adds an important point: the earlier configuration of a stratum now sandstone is not actually observed but is inferred from the later. The examples illustrate two kinds of explanatory sequences available to the historical scientist. In one we have dated documents contemporaneous with the events and so directly historical in nature and sequence. In the other we have a pseudohistorical sequence such as that of presently existing sands and sandstones. Their resemblances and differences are such that we can be confident that they share some elements of historical change, but that one has undergone more change than the other. In this case it is easy to see that the sandstone belongs later in the pseudohistorical sequence. One therefore infers for it a historically antecedent sand and can proceed to determine what characteristics are inherited from that sand and the nature of the subsequent changes.

Pseudohistorical sequences are most extensively used in historical biology, where historical inference must so often be based on living organisms that are in fact contemporaneous and not sequential. The basis of this comparative method for attacking historical problems is that it commonly occurs that certain characteristics have evolved more slowly in one lineage than in another. Although the principle is so simple, its application is often difficult and replete with puzzles. The nature and direction of the sequence is almost never as simple as in the inorganic sand–sandstone example.

A second form of strategy has a certain analogy with the use of multiple experiments with controlled variables. The method is to compare different sequences, either historical or pseudohistorical, that resemble each other in some pertinent way. Resemblances in the antecedent configurations may be taken to include causes of the consequent resemblances. It is not, however, legitimate to assume that they are all necessary causes or that they include sufficient causes. Even more important at times is the converse principle that factors different among the antecedents are not causes of resemblances among the consequents. By elimination when many sequences are compared, this may warrant the conclusion that residual antecedent resemblances are necessary causes. There is here applicable a principle of scientific testing in general: absolute proof of a hypothesis or other form of inference is impossible, but disproof is possible. Confidence increases with the number of opportunities for disproof that have not in fact revealed discrepancies. In this application, confidence that residual resemblances are causal increases with the number of different sequences involved in the comparison. This form of strategy is applicable to most evolutionary sequences, few of which are unique in *all* respects.

An interesting special case arises when there is more resemblance among consequent than among antecedent configurations, the phenomenon of convergence. This has received much more attention in the study of organic evolution than elsewhere, but nonorganic examples also occur. In organic evolution it has greatly increased understanding of the nature and limits of adaptation by natural selection.

It has been previously pointed out that the explanation of a historical event involves both configuration and immanence, even though the latter is not historical in nature. Historical science therefore requires knowledge of the pertinent immanent factors and its strategy includes distinguishing the two and studying their interactions. Nonhistorical science is particularly (although not exclusively) concerned with the immanent. It is the principal source of the historian's necessary knowledge of immanent factors and the principal means of distinguishing them from configurational relationships. A typical approach is to vary configurations in experiments and to determine what relationships are constant throughout the configurational variations. To a historical biologist, the function of biochemists and biophysicists is to isolate and characterize the immanent properties of the constituents of cells in that and other ways. The historical biologist is then interested not in what holds true regardless of configuration, but in how configuration modifies the action of the identified immanent properties and forces. In this respect, the nonhistorical scientist is more interested in similarities and the historical scientist in differences.

Here the historical scientist has two main strategies, both already mentioned. They may be used separately or together. One is by controlled experimentation. The other might be viewed as complementary to the previously discussed study of similarities in multiple sequences. In this strategy, attention is focused on consequent differences, the causes of which are sought among the observable or inferable differences of antecedent configurations.

Points always at issue in historical science are the consistency of proposed immanent laws and properties with known historical events and the sufficiency and necessity of such causation acting within known configurations. Probably the strongest argument of the catastrophists was that known features of the earth were inconsistent with their formation by known natural forces within the earth's span of existence, which many of them took to be about 6000 years. The fault of course was not with their logic but with one of their premises. The same argument, with the same fallacy, was brought up against Darwin when it was claimed that his theory was inadequate to account for the origin of present organic diversity in the earth's span, then estimated by the most eminent physicists as a few million years at most. Darwin stuck to his guns and insisted, correctly, that the calculation of the age of the earth must be wrong. Historical science has an essential role, both philosophical and practical, in providing such cross-checks (mostly nonpredictive and nonexperimental), both with its own theories and with those of other sciences as part of the self-testing of science in general.

The testing of hypothetical generalizations or proposed explanations against a historical record has some of the aspects of predictive testing. Here, however, one does not say, "If so and so holds good, such and such will occur," but, "If so and so has held good, such and such must have occurred." (Again I think that the difference in tense is logically significant and that a parity principle is not applicable.) In evolutionary studies a conspicuous example has been the theory of orthogenesis, which in the most common of its many forms maintains that once an evolutionary trend begins it is inherently forced to continue to the physically possible limit regardless of other circumstances. That plainly has consequences that would be reflected in the fossil record. As a matter of observation, the theory is inconsistent with that record.

The study of human history is potentially included in historical science by our definition. One of its differences from other branches of historical science is that it deals with configurational sequences and causal complexes so exceedingly intricate that their scientific analysis has not yet been conspicuously successful. (Toynbee's correlation of similar sequences would seem to be a promising application of a general historical strategy, but I understand that the results have not been universally acclaimed by his colleagues.) A second important difference is that so much of this brief history has been directly observed, although with varying degrees of accuracy and acuity and only in its very latest parts by anyone whose approach can reasonably be called scientific. Direct observation of historical events is also possible in biology and other historical sciences, and it is another of their important strategies.

In biology, however, and in all historical science except that of human history, the strategic value of observing actual events is more often indirect than direct. The processes observed are, as a rule, only those that act rapidly. The time involved is infinitesimal in comparison with the time span of nonhuman history, which is on the order of $n10^9$ *years* for both historical geology and historical biology. Directly observed events are also both local and trivial in the great majority of instances. They are in fact insignificant in themselves, but they are extremely significant as samples or paradigms, being sequences seen in action and with all their elements and surrounding circumstances observable. They thus serve in a special and particularly valuable way both as historical (and not pseudohistorical) data for the strategies of comparison of multiple sequences and as natural experiments for the strategies of experimentation, including on some but not all occasions that of prediction.

Direct observation of historical events is also involved in a different way in still another of the historical strategies, that of testing explanatory theories against a record. For example, such observations are one of the best ways to estimate rates of processes under natural conditions and so to judge whether they could in fact have caused changes indicated by the record in the time involved. Or historical importance of observed short-range processes can be tested for necessity, sufficiency, or both against the long-range record. An interesting paleontological example concerns the claims of some Neo-Lamarckians that although the inheritance of acquired characters is too slow to be directly observed it has been a (or the) effective long-range process of evolution. The fossil record in itself cannot offer clear disproof, but it strips the argument of all conviction by showing that actually observed short-range processes excluded by this hypothesis are both necessary and sufficient to account for known history.

The most frequent operations in historical science are not based on the observation of causal sequences—events—but on the observation of results. From those results an attempt is made to infer previous causes. This is true even when a historical sequence, for example one of fossils, is observed. Such a sequence is directly historical only in the sense that the fossils lived in a time sequence that is directly available to us. The actual events, the lives of the animals and their burial in the rocks, are not observable. In such situations and in this sense the present is not merely a key to the past: it is all we have in the way of data. Prediction is inferring results from causes. Historical science is largely involved with quite the opposite: inferring causes (of course including causal configurations) from results.

The reverse of prediction has been called postdiction. In momentary return to the parity of explanation and prediction, it may be noted that if A is the necessary and sufficient cause of X and X is the necessary and sole result of A, then the prediction of X from A and the postdiction of A from X are merely different statements of the same relationship. They are logically identical. It has already been demonstrated and sufficiently emphasized that the conditions for that identity frequently do not hold in practice and sometimes not in principle for historical science. Here, then, postdiction takes on a broader and distinct meaning and is not merely a restatement of a predictive relationship. With considerable oversimplification it might be said that historical science is mainly postdictive and non-historical science mainly predictive.

Postdiction also involves the self-testing essential to a true science, as has also been exemplified although not, by far, fully expounded. Perhaps its simplest and yet most conclusive test is the confrontation of theoretical explanation with historical evidence. A crucial historical fact or event may be deduced from a theory and search may subsequently produce evidence for or against its actual prior occurrence. That has been called "prediction," for example by Rensch, sometimes with the implication that historical science really is science because its philosophical basis does not really differ from that of nonhistorical physics. The premise that the philosophy of science is necessarily nonhistorical is of course wrong, but the argument is fallacious in any case. What is actually predicted is not the antecedent occurrence but the subsequent discovery; the antecedent is postdicted. Beyond that perhaps quibbling point, the antecedent occurrence is not always a *necessary* consequence of any fact, principle, hypothesis, theory, law, or postulate advanced before the postdiction was made. The point is sufficiently illustrated on the pragmatic level by the sometimes spectacular failure to predict discoveries even when there is a sound basis for such prediction. An evolutionary example is the failure to predict discovery of a "missing link" now known (*Australopithecus*) that was upright and tool-making but had the physiognomy and cranial capacity of an ape. Fortunately such examples do not invalidate the effectiveness of postdiction in the sense of inferring the past from the present with accompanying testing by historical methods. In fact that discovery was an example of such testing, for without any predictive element it confirmed (i.e., strengthened confidence in) certain prior theories as to human origins and relationships and permitted their refinement.

Another oversimplified and yet generally significant distinction is that historical science is primarily concerned with configuration and nonhistorical science with immanence. Parallel, not identical, with this is a certain tendency for one to concentrate on the real and the individual, the other on the ideal and the generalized, or for them to operate with different degrees of abstraction. We have seen, however, that interpretation and explanation in historical science *include immanence* and along with it *all* the facts, principles, laws, and so on of nonhistorical science. To these historical science adds its own configurational and other aspects. When it is most being itself, it is compositionist rather than reductionist, examining the involvement of primary materials and forces in systems of increasing complexity and integration.

Historical science, thus characterized, cuts across the traditional lines between the various sciences, physics, chemistry, astronomy, geology, biology, anthropology, psychology, sociology, and the rest. Each of these has both historical and nonhistorical aspects, although the proportions of the two differ greatly. Among the sciences named, the historical element is least in physics, where it is frequently ignored, and greatest in sociology, where the existence of nonhistorical aspects is sometimes denied—one of the reasons why sociology has not always been ranked as a science. It is not a coincidence that there is a correlation with complexity and levels of integration, physics being the simplest and sociology the most complex science in this partial list. Unfortunately philosophers of science have tended to concentrate on one end of this spectrum, and that the simplest, so much as to give a distorted, in some instances quite false, idea of the philosophy of science as a whole.

PART THREE

The Problem of Purpose

CHAPTER EIGHT

The History of Life

IN the next few chapters I shall discuss some of the major philosophical issues that arise from the fact of evolution and from the history of life. Much of the basis for that discussion has already been laid. The fact of evolution, its main processes or mechanisms, and the nature of history have been considered, among other topics. The course of the history of life has not been. The purpose of the present chapter is to discuss that course, not by following it in a descriptive narrative but by investigating the *sort* of thing it is, the main characteristics of organic history in its broad sweep.

If we had no direct record of the history of life, three alternatives might seem most logical. Most obvious would be the supposition that seems to be exemplified in the world we see, the only one of which we have any human record: a dynamic equilibrium with individuals being born and dying, populations waxing and waning, but no apparent over-all trend. That is, of course, essentially the creationist dogma, but it could also be an outcome of evolution. With realization of the fact of evolution, two other possibilities (both of which were more dimly involved in much earlier speculation) become clearer. There might be a steady progression step by step from, as the saying goes, ameba to man, with each earlier and lower stage remaining static or, at least, in stabilized equilibrium after the next higher stage arose from it. Or there might be a steady expansion and fragmentation of the world of life as organisms occupied all environments and different forms became increasingly specialized for particular niches within those environments.

The last two alternatives are not mutually exclusive, and both ideas have always been and indeed still are simultaneously involved in most thinking about evolution. They do not, moreover, exclude the first alternative, for it would seem that neither progression nor expansion can go on forever and that an equilibrium must at last be reached. Opinions differ as to whether it has already been reached. The record now acquired plainly shows that all three processes are indeed intricately interwoven in the fabric of history but that the pattern is even more complicated than any intermixture of those three motifs. Another equally important element occurs: extinction.

It seems curious now that savants ever argued as to whether species can become extinct. Thomas Jefferson on religious grounds and Lamarck for more scientific reasons agreed that extinction had not occurred, even while Cuvier was demonstrating that it had. A priori there is no strictly logical necessity for it. Why should progression and expansion not simply fill the world with living things that would then preserve their status? Only the fossil record could establish the really startling fact that extinction is the *usual* fate of species and that it has made history quite different from what once seemed logical to Lamarck and many other logical men.

The fossil record suffices to confirm that the history of life has involved, in the main, these four grand processes: expansion, progression, equilibrium (or stabilization), and extinction. No one of these has been constant or dominant over-all. No one characterizes all organisms at any given time or all times for any given organisms. Moreover, they are never independent, but each process depends on and in turn helps to determine all the others. The result is extremely complex, but it can be analyzed in its broader and some of its minor lines, at least in a descriptive way. Causal analysis is, as usual, more difficult and less secure, but it is possible, and herein are the deepest meanings and the highest rewards of the study of the history of life.

It is probable that life originated in only a few kinds of organisms, perhaps, but not necessarily, only one. The early history of organisms must then have been especially characterized by expansion, by proliferation both of individuals and of kinds. The known fossil record gives virtually no information about that earliest expansion. It is extremely improbable that the first living things left recognizable traces. There are probable fossil remains of organisms that are close to two billion years old, but beyond the fact of their existence these are not very informative. Thereafter, as one would expect, known fossils do become clearer, less rare, and more varied, but until the beginning of the Cambrian they are still too few to constitute a really useful historical record.

From the Cambrian onward there is such a record. (If the names of geological eras, periods, and epochs used here are unfamiliar, consult the table given in the notes to this chapter at the end of the book.) The Cambrian record picks up what is clearly a major expansion in progress. That expansion continued for a very long time, probably well over a hundred million years, into the middle of the next geological period, the Ordovician.

Table 1 shows a clear increase in number of known major groups (phyla and classes) from Early Cambrian to Middle Ordovician. When comparison is made with Recent marine animals that would probably occur in the earlier record if they were living at the time, the number of phyla is now the same as in the Middle Ordovician, and the number of classes is actually slightly smaller. Numerous relatively unfossilizable soft-bodied groups—those absent or extremely rare in any part of the fossil record—are omitted from the Recent count, but there is no reason to believe that their trend in diversity was notably different from that of the groups recorded as fossils. (Indeed, these less fossilizable groups may well have been relatively *more* common in the Cambrian.) Comparison of lower hierarchic categories, especially of genera and species, is too uncertain to be significant; we can be sure that at these levels the Cambrian and Ordovician record is highly incomplete but cannot usefully estimate just how incomplete.

Table 1   Numbers of Marine Animal Phyla and Classes in Cambrian, Ordovician, and Recent

Time	Phyla	Classes
CAMBRIAN:		
Early	8	12
Middle	10	20
Late	11	22
ORDOVICIAN:		
Early	11	27
Middle	12	33
Late	12	33
RECENT	12	31

It is concluded that basically, at the level of classes and above, the diversity of the marine fauna increased into the Middle Ordovician but has not increased since then. Doubtless it has fluctuated, and this must notably have been true of lower taxonomic levels, but the major expansion of marine animals was evidently complete in the Ordovician and has been more or less in equilibrium since then. The living phyla are the same as those of the Ordovician, and, at this most fundamental level of all, the subsequent equilibrium has been static. Within the phyla, however, the equilibrium has been quite dynamic. Several of the classes and an increasing proportion of the lesser groups have become extinct and have been replaced by groups of separate origin. The proportion of Ordovician species that have living descendants is unknown but is certainly minute.

The maintenance of such dynamic equilibrium involves repeated contractions and expansions within the pattern even while the over-all diversity remains roughly constant within broad limits.

There is a relay effect, and this is perhaps the element we would least expect if we did not have the fossil record to demonstrate it. A broad ecological zone may remain approximately filled for long periods of time, but the groups actually occupying it may change radically, one relaying the other. Thus the diversity of hoofed herbivorous mammals changed only within broad limits from about Late Paleocene into the Pleistocene, but the predominant groups were first various archaic orders, then perissodactyls, and finally artiodactyls. Data for North America are given in Table 2. Here there has been a marked decline in total diversity since the Miocene, a complex phenomenon probably due to a variety of factors: relay by nonungulates, increasing dominance of a few exceptionally successful species, floral and climatic changes, and probably other influences. However that may be, the relay phenomenon within this broad ecological type is evident.

The example shows that relaying may occur during times of expansion, equilibrium, and decline of numbers of taxa and is not confined to equilibrium phases. On a still broader scale, relay during expansion is illustrated by the land vertebrates. With fluctua-

Table 2   Relay in North American Ungulates: Percentage of Genera Belonging to Various Orders in Cenozoic Epochs

	Archaic Orders*	Proboscidea	Perisso-dactyla	Artiodactyla	Total No.
RECENT	0	0	0	100	10
PLEISTOCENE	0	11	11	78	27
PLIOCENE	0	20	20	61	46
MIOCENE	0	6	22	72	69
OLIGOCENE	0	0	38	62	47
EOCENE	14	0	52	33	84
PALEOCENE	100	0	0	0	36

* Condylarths, pantodonts, uintatheres, and one notoungulate.

tions and shorter times of stasis and moderate contraction, that expansion continued from the Devonian well into the Cenozoic, but the main expansive activity was at first among amphibians, then reptiles, and finally mammals.

The last example is part of the second major over-all expansion of animals shown in the record, following the marine expansion seen in the Cambrian and Ordovician. It reflects spread into another major environment or complex of environments—that of the land. That began slowly, as far as the record shows, in the Silurian, was in full swing in the Carboniferous, and continued with fluctuations and some setbacks well into the Cenozoic. No new phylum was involved and only a few of the older, previously aquatic phyla (hardly any forms outside the Mollusca, Arthropoda, and Chordata), and a modest number of classes, but the terrestrial fauna became extremely diverse at the levels from orders to species.

Those features of the record and some others are reflected in Table 3. The exact numbers there given have little significance, since no two tabulators would agree exactly as to which groups are "well-recorded" or would use just the same classification. Certain general features seem, however, to be reliable. All sufficiently known phyla and an absolute majority of known classes first appear in the Cambrian and Ordovician. No phylum is known to have arisen since then, and only a few classes are known to have arisen since the Carboniferous—only one (Mammalia) is tabulated; at least one more (Aves) is known to have originated after the Carboniferous, and a few more may have.

Basic stabilization of marine invertebrate faunas well before the end of the Paleozoic is evident at the level of phyla and classes. The orders, especially, also reflect approach to stabilization in the Ordovician, but there is another proliferation of orders in the Carboniferous and a sequence of them through the Mesozoic. Those for the most part represent relay phenomena, new groups replacing old in a fluctuating dynamic equilibrium. The marine vertebrates expanded and stabilized much later than did the invertebrates. This was in part a relay of invertebrates by vertebrates. The aquatic vertebrates reached a near-equilibrium in both classes and orders by

Table 3   *Times of First Appearance in Fossil Record of Well-recorded Groups of Animals*

Time	Invertebrates,* Mostly Marine			Vertebrates†				Totals		
				Fishes		Tetrapods				
	Phyla	Classes	Orders	Classes	Orders	Classes	Orders	Phyla	Classes	Orders
CENOZOIC	0	0	4	0	—‡	0	25	0	0	29‡
CRETACEOUS	0	0	10	0	5	0	3	0	0	18
JURASSIC	0	0	11	0	5	1	8	0	1	24
TRIASSIC	0	0	9	0	3	0	10	0	0	22
PERMIAN	0	0	1	0	1	0	3	0	0	5
CARBONIFEROUS	0	3	22	0	0	1	9	0	4	31
DEVONIAN	0	6	9	2	11	1	1	0	9	21
SILURIAN	0	3	6	1	5	0	0	0	4	11
ORDOVICIAN	1	14	46	1	1	0	0	2§	15	47
CAMBRIAN	6	21	32	0	0	0	0	6	21	32

* Omitting insects and other groups with poor records.
† Omitting birds and lesser groups with poor records.
‡ Many orders of teleosts appear with fair records in the Cenozoic, but their systematics is so poorly established that they are omitted. This makes the total for Cenozoic orders relatively too small.
§ The second phylum, additional to that in the first column, is Chordata.

the end of the Devonian. Renewed proliferation of orders in the Middle to Late Mesozoic and (not shown by the tabulation but probable on other data) Early Cenozoic involves relays and probably also a renewed expansion (both involving almost solely the teleosts).

Proliferation of land vertebrates approached an early and temporary equilibrium in the Carboniferous, mainly among the imperfectly terrestrial amphibians. A relay plus a definite new expansion into fully terrestrial habitats occurred among the reptiles, especially in the Jurassic, and is evident in the table at the ordinal level. Another relay and still further expansion involve the Cenozoic mammals. Although not tabulated, there was also a great expansion of birds mainly in the latest Mesozoic and Early Cenozoic. This was in small part a relay (replacing pterosaurs), but it was mainly an expansion into a new set of ecological situations, most of which were previously empty or nonexistent. A concomitant great expansion of mainly volant nonaquatic invertebrates (especially insects) also occurred and probably was not stabilized until well into the Cenozoic. The record is still too poor for reliable tabulation, but the broad event is evident. Both these expansions and some others correlate with Late Mesozoic expansion of angiospermous plants, which created innumerable previously nonexistent niches for animals.

The relays and post-Ordovician expansions of special groups reached up to the ordinal level and occurred at various times including the Cenozoic. For those reasons there is no apparent over-all diminution in the rise of new orders of animals, from Cambrian to Cenozoic, in sharp contrast to the situation as regards classes and phyla. That contrast would be still more striking if reliable figures were at hand for three groups omitted from the present tabulation and particularly rich in orders of comparatively late origin: insects, teleost fishes, and birds. There are, of course, fluctuations, notably lows in the Silurian and Permian which are probably original phenomena and not sampling errors.

Darwin tells us in his *Autobiography:*

At that time [1844] I overlooked one problem of great importance. . . . This problem is the tendency in organic beings descended from the same stock to diverge in character as they become modified. . . . I can remem-

ber the very spot in the road . . . when to my joy the solution came to me.
. . . The solution, as I believe, is that the modified offspring of all dominant and increasing forms tend to become adapted to many and highly
diversified places in the economy of nature.

That is, of course, the principle of adaptive radiation, the "discovery" of which has enhanced several reputations since Darwin.

Darwin, in *The Origin of Species,* also perceived a relationship
between the origin of new and the extinction of older groups. He
concluded that, in later geological times at least, there was what has
here been called a dynamic equilibrium and that "the production
of new forms has caused the extinction of about the same number of
old forms." He considered that the forms thus becoming extinct
would, not invariably but as a rule, be those most closely related to
the new forms.

'I'hose processes envisaged by Darwin are confirmed by our
greater knowledge of the paleontological record. Among the details
of the history is a continual replacement of species by related species
and genera by related genera. On a broader scale we now see, even
more clearly than Darwin did, that every marked expansion of a
group, whether it be a genus or a phylum or the whole animal kingdom, is an adaptive radiation. Each starts with a group of a certain
adaptive status in a particular range of environments. (Such a group
is "generalized" only in the sense that it is not irrevocably committed to a special, narrow range.) Each radiates by a combination
of two processes: a parceling-out of a broader ecological range
among more specifically adapted separate lines of descent and the
invasion of new ecological niches by modifications of the ancestral
adaptation.

From what has already been said, it is evident that there are
primary expansions and relaying expansions, the patterns within
both kinds of expansions normally being those of adaptive radiation. Primary expansion represents the occupation of hitherto empty
ecological situations, as in the occupation of the seas from the Pre-
cambrian into the Ordovician and the later occupation of the lands.
The scale need not be so grand or the events so ancient. The relatively recent occupation of the Galápagos Islands by finches was a

primary radiation. In such events extinction may occur, especially on the small scale of species expanding at the expense of closely related species, as stressed by Darwin; but here extinction is almost irrelevant to the main phenomenon.

A relay expansion is a reoccupation, frequently or even typically, by a group that is *not* closely related to the one replaced. Extinction is an essential part of this phenomenon, for relay does not occur unless the older group is reduced in scope or entirely eliminated. In many, probably most, instances the contraction of one group and expansion of another are concomitant and are almost certainly causally related, that is, the spread of the relaying group is itself the cause or one of the causes of extinction. A well-documented example is the radiation of invading Pliocene-Pleistocene North American mammals in South America and the concomitant extinction of many old natives of South America. Another striking example has gone on before our eyes as the expanding introduced placentals have caused restriction of marsupials in Australia, but there the invasion is so recent that a true adaptive radiation of the invaders has not had time to occur. It will presently be noted that one group may be relayed by several. It is also true that several groups may be relayed by one, as has clearly happened in Australia and probably also among North American ungulates since the Miocene (see Table 2).

In other, more puzzling instances extinction seems definitely to have preceded the relaying expansion, and the causal relationship is apparently reversed: new radiation occurs because of prior extinction, instead of extinction resulting from the radiation. There are, as Darwin also noted but did not stress, causes of extinction other than competitive replacement. The most striking example is the relaying of most of the Mesozoic reptiles by Cenozoic mammals, with no evidence of competition.

Another point well exemplified by the Mesozoic reptiles versus the Cenozoic mammals is that a replacing group may be able to exploit an ecological situation more intensely and to subdivide it more finely than the older group. The Cenozoic mammals do cover roughly the same ecological range as older reptiles, but they are far more diverse, with many more different specific adaptations. Similar phenomena have occurred with some frequency in other groups, for instance in fishes with the expansion of the teleosts.

In saying that animals of a given sort of adaptation are, if they become extinct, replaced or relayed by others, there is no implication that the later forms will be structurally similar. It suffices that the food, the habitat, in general the *Lebensraum* of an earlier group be utilized by a later one. The replacers may be very different in size and other characters, and one group may be replaced by several others, each partly occupying the previous situation. There are no mammals really at all like any dinosaurs, but the dinosaurs' places are thoroughly taken by diverse mammals. This is another conclusion from the record that could hardly be grasped in Darwin's day. He followed Buckland (a nonevolutionist) in thinking that "extinct species can all be classed either in still existing groups, or between them." That is true only at levels so broad that the statement is almost meaningless or in a still quite broad ecological and not phylogenetic sense. Surely no one would now claim, for instance, that the dinosaurs (a few of which were known to Darwin) belong in any meaningful way either to or between still existing groups, even though there are living members (the crocodiles) of the same subclass. A less familiar but even more extreme example is provided by the graptolites, a once abundant, long extinct group so unlike any Recent animals that it is doubtful what phylum they belong to, if, indeed, they should not be placed in a phylum of their own.

Innumerable extinct groups have been relayed by organisms markedly different from those replaced. Almost all ancient radiations of considerable scope include some extinct groups strangely different from any later groups, even though some later groups do seem in most cases to have occupied their ecological positions. Examples include the hippurites among mollusks, dicynodonts among reptiles, diatrymids among birds, uintatheres among mammals, and many others. Only in some rather late radiations that still have not definitely passed their acme, do strikingly unique extinct groups seem to be absent—for instance, in Late Mesozoic and Cenozoic angiosperms, insects, or teleosts. And even in them the apparent absence of (to recent eyes) really queer extinct groups may be due to ignorance.

Here the question arises whether certain extinct groups have in fact been relayed or whether, as environments have evolved, the ecological opportunities exploited by those groups have not simply ceased to exist. An unequivocal answer is hardly possible. There is no real doubt that relaying has often occurred, and the impression is that it is usual. The fact that extinct groups are often so different from the survivors makes it impossible in some instances to be explicit as to the identities of possibly relaying groups. The ecology of graptolites, for instance, is not clear enough to identify relaying groups precisely, but there are plenty of possibilities: for the sessile forms perhaps various bryozoans and coelenterates; for the planktonic graptolites various protozoans and small arthropods.

Relaying, as between dinosaurs and mammals, apparently can be delayed and so might be delayed indefinitely. Environments do change, and so some must wholly disappear. Yet I can think of few extinct groups of which it can confidently be said that relay has not occurred. A fairly typical example of the *possibility* is provided by the chalicotheres, an extinct group of large, clawed ungulates. Their way of life has been the subject of some wild surmises, but it is really unknown. There is nothing at all like them today, so perhaps their ecological niche is now empty or no longer exists, but how can one decide? It might be too close to a circular argument to reason that the niche no longer exists from the fact that the chalicotheres became extinct or from the "fact" (not really known to be a fact) that there is no ecological replacement for them.

There is a widespread impression that radiations have tended to become smaller in scope (cover smaller ecological ranges and involve lower taxonomic levels) in the course of geological time. That doubtless has a measure of truth, but it requires qualification. Primary radiations, although they can be of any extent, can be the largest of all, and a really large primary radiation can occur only once. Thereafter a new radiation in the same ecological field will necessarily be a relay. The last major primary radiations were in the occupation of land and air, essentially completed in the Mesozoic.

The surprising thing is not that it was so early but that it was so late, at least 1500 million years after the origin of life! Relaying radiations do not have such wide possible scope and characteristically occur, in taxonomic terms, at the level of orders and below. The data of Table 3 suggest that there has been no over-all tendency for them to decrease in either number or scope since the Cambrian. None of ordinal scope has occurred since the Early Cenozoic, but that may be merely because it takes a great deal of time to develop that scope. A radiation that started in the later Cenozoic would not yet have reached a degree of divergence designated as ordinal in current taxonomy.

Although relaying radiations have not tended to decrease regularly, they are not randomly distributed in time. There have been several definite times when major extinctions, up to about the ordinal level, have been especially prevalent, and, as would be expected, these are accompanied by the beginnings and followed by the peaks of major relaying radiations. The first such episode for animals (after the Cambrian) has extinctions distributed around the Silurian-Devonian boundary, with a complex of relaying radiations in the Devonian and into the Carboniferous, where that expansion reaches its acme. Those relays, as now known, are all in marine environments. Insects and amphibians appear in the Devonian, reptiles in the Carboniferous. Those are parts of a primary radiation, but the coincidence with intense relaying radiation may be significant. The next major episode of extinction-then-radiation is in the late Permian and the Triassic, again mainly marine but also affecting some terrestrial groups. The last on this major scale, mainly terrestrial but also in part marine, was the famous Late Cretaceous–Early Cenozoic changeover. Similar episodes of shorter duration and less intensity also occur, for instance, in Late Triassic–Early Jurassic (extinction of many archaic reptiles, origin of mammals, almost complete relay among ammonites, and so on).

The division of later geological time into Paleozoic, Mesozoic, and Cenozoic is based on the major extinctions-relays of the Permian-Triassic and Cretaceous-Tertiary. The Paleozoic could well have been divided into two eras at the Silurian-Devonian boundary, but that episode is not quite so striking and was not so well known to early geologists. It has often been noted that the history of land plants would support a different division of "Paleophytic," "Mesophytic," and "Cenophytic." The primary radiation was mostly Devonian, when there was much relaying and some primary radiation among animals, and the first great extinction-then-relay was Permian-Triassic, coinciding with a similar episode among animals. The last major episode of that sort among land plants was, however, Jurassic-Cretaceous, and there was no really important change between Mesozoic and Cenozoic.

The Triassic-Jurassic episode among animals, although unknown to Darwin, strikingly illustrates one of his points: "To feel no surprise at the rarity of a species, and yet to marvel greatly when the species ceases to exist, is much the same as . . . to feel no surprise at sickness, but, when the sick man dies, to wonder and to suspect that he died by some deed of violence." Only one family (Phylloceratidae), perhaps even only one genus (*Phylloceras* or its immediate ancestor), of ammonites survived the Triassic, but from that (one might think almost chance) survival the group became tremendously diverse by relay. The total extinction of ammonites around the end of the Cretaceous differs only in that no genus happened to survive. The event is no more wonderful than the near-extinction at the end of the Triassic and need have no other cause —whatever that wholly unknown cause may be!

Another of Darwin's opinions about extinction is not completely substantiated by better knowledge of the record: "that the extinction of a whole group of species is generally a slower process than their production." That is frequently true, but the exact opposite is also frequently true. (Darwin noted that there are exceptions to his rule, citing the Cretaceous ammonites.) There is great diversity in patterns of expansion and contraction and a general rule can hardly be formulated.

The history of life is decidedly nonrandom. This is evident in many features of the record, including such points already discussed as the phenomena of relays and of major replacements at defined times. It is, however, still more striking in two other phenomena copiously documented by fossils. Both have to do with evolutionary trends: first, that the direction of morphological (hence also functional and behavioral) change in a given lineage often continues without significant deviation for long periods of time and, second, that similar or parallel trends often appear either simultaneously or successively in numerous different, usually related, lineages. These phenomena are far from universal; they are not "laws" of evolution; but they are so common and so thoroughly established by concrete evidence that they demand a definite, effective directional force among the evolutionary processes. They rule out any theory of purely random evolution, such as the rather naïve mutationism that had considerable support earlier in the twentieth century. What directional forces the data do demand, or permit, is one of the most important questions to be asked of the fossil record.

An important point about trends in particular lineages is that they must be isolated and studied carefully in true temporal sequences, not accepted as vague generalizations or arranged subjectively from scattered individual observations. Perhaps the most widely known example is the evolution of *Equus* from "little eohippus," frequently represented as a steady "'orthogenetic" progression. Examined with care, it is nothing of the sort. Size, for instance, was approximately stable for the first 20 million years or so, then tended to increase on an average, although from time to time various branches stopped becoming larger or perhaps even became smaller. Different lineages progressed at very unequal rates, so that at any one time in the later Tertiary there were horses of extremely different sizes, some larger than any living horse and some at least as small as their Oligocene forebears. Molarization of the premolars was rather steadily progressive for perhaps 25 million years and then ceased to evolve in all lines. Change in number of toes and foot structure was never a steady trend but was transformed quite rapidly three times, and each of the last two times saw various lineages that were *not* affected and remained on the evolutionary level reached earlier. (This example also illustrates another important phenomenon that cannot here be further discussed: adaptive stabilization.)

Another famous example, that of the supposed "orthogenetic" increase in size of canines in sabertooths (or "sabertoothed tigers," but they were not tigers), is apparently spurious. It can be supported only by a subjective arrangement of animals that were not in the same or closely related lineages and are not taken in their actual temporal sequence. As far as real analysis has yet gone, no tendency toward increase in size of those teeth is evident among truly related animals in their real sequence.

It is not the point that trends are really nonexistent; they certainly are real and frequent, but the doctrine of "orthogenesis" has grossly exaggerated their frequency, duration, and continuity. They may be absent in a given phylogeny (except as any change has *some* extent of duration and *some* direction). They start and stop, sometimes apparently erratically. They are occasionally reversed. They do not endure indefinitely, the usual order of magnitude being some 10^7 years and in real, established examples rarely, if ever, longer than about 5×10^7 years at anything like constant rate and direction. Even in one lineage, different structural changes commonly occur at different times and different rates. Progressive changes often occur in a series of fairly abrupt (not, however, instantaneous as far as can be established) steps and not in a continuous trend. Comparatively few paleontologists now accept the theory of orthogenesis, as an inherent tendency for evolution to proceed indefinitely on a certain route, once embarked on it. The record just does not support that generalization.

The phenomenon of parallel trends may be illustrated by a particularly striking, large-scale example. The living mammals, monotremes excepted, are clearly of monophyletic derivation from reptiles. Long ago I suggested that the known Mesozoic mammals are not. That idea has been strongly supported by later studies and now seems well established. At least six different lineages probably crossed the conventional line providing the usual structural distinction between reptiles and mammals at about the same time and each one independently: tritylodonts, multituberculates, triconodonts. symmetrodonts, docodonts, and pantotheres. The marsupials and placentals are derived from the pantotheres (*sensu lato*, at least). The monotremes may be derived from one of the other Mesozoic orders or may represent still another separate crossing of the line.

On the reptilian side, numerous different lineages of the order Therapsida were independently acquiring various mammal-like characters. Some of the evidence is summarized in Table 4. The full situation was much more striking and complex, for other groups of

Table 4 Independent, Progressive Acquisition of Mammal-like Characters by Various Groups of Therapsid Reptiles

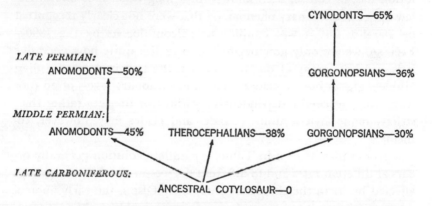

EARLY TO MIDDLE TRIASSIC:

CYNODONTS—65%

LATE PERMIAN:
ANOMODONTS—50%   GORGONOPSIANS—36%

MIDDLE PERMIAN:
ANOMODONTS—45%   THEROCEPHALIANS—38%   GORGONOPSIANS—30%

LATE CARBONIFEROUS:
ANCESTRAL COTYLOSAUR—0

Independent, progressive acquisition of mammal like characters by reptilian forerunners of early mammals. As is usual in geological and phylogenetic diagrams, time is shown as proceeding from bottom to top of the diagram. The groups named are those of mammal-like reptiles, called "therapsids" as a whole, and of the most primitive, ancestral reptiles: the cotylosaurs. Arrows indicate lines of phylogenetic descent and progressive evolution. The percentage figures indicate how far, at the given time, a group has progressed from the primitive reptilian condition—zero per cent—toward a fully mammalian condition—100 per cent. The figures show that each line of descent was independently becoming more and more mammal-like, although the rate of progress was different in the different lines. Various later lines (not shown here) had reached almost fully mammalian status (85–95 per cent) in the following period, the Jurassic, and some were completely mammalian (100 per cent) in the next period after that, the Cretaceous. (Based on data from Olson, 1944) therapsids were also evolving more or less in parallel with these, and within all groups there were many distinct lineages with considerable parallel advance toward mammalian structure in various ways. In the face of all this evidence it is no longer possible to accept the earlier argument that a change so intricate as incorporation of bones of the lower jaw into the ear could only have happened once. It happened several or many times.

That example is exceptional both in taxonomic level and in the magnitude of the changes involved. The phenomenon that it represents is, however, extremely common. The existence of parallel evolution has, of course, been known for a long time. It is one of the few major evolutionary phenomena that were not clearly recognized by Darwin, but it was familiar to paleontologists by the 1890's. Even so, we are only now beginning to realize quite how pervasive it is and how many of our taxa are, in the terminology of Julian Huxley, grades but no clades, that is, evolutionary levels often (not invariably) reached independently by different lineages rather than strictly monophyletic units. (Grades and clades may, however, coincide.)

As can also be seen in Table 4, parallel evolution generally occurs at different rates and to different degrees in the different groups affected by it. In the characters studied by Olson, the early anomodonts (=dicynodonts) were far more mammal-like than contemporaneous gorgonopsians, but the anomodonts became extinct long before reaching the structural reptile-mammal line, while some probable descendants of the gorgonopsians crossed that line. Another essential point is that parallelism involves only certain characters, while other characters in the same animals may evolve in quite different directions in different lineages. While the anomodonts and gorgonopsians were both independently acquiring some of the same mammal-like features, they were also each acquiring their own peculiar characteristics, so that late anomodonts and gorgonopsians were very different indeed. Also the six groups previously mentioned as having separately crossed the arbitrary reptile-mammal line are radically dissimilar in most ways except those that place them all by definition in the grade (not clade) Mammalia. It is, then, necessary in such episodes to distinguish clearly between the features involved (at different rates and in different degrees) in parallel evolution and those simultaneously involved in adaptive radiation.

What are the possible theoretical or causal explanations of these phenomena? The principal alternatives that have adherents today are natural selection, directional mutation, and a vitalistic or perfectionistic inner urge or inherent tendency. The selectionist theory is that a trend is adaptive for the lineage involved, that it continues only as long as it is adaptive, that it stops when adaptation is as complete as selection can make it in given circumstances, and that it changes or the group becomes extinct if a different direction of evolution becomes adaptive. Often the adaptive nature of a trend seems apparent. Often it is not apparent, but the postulate still seems required to account for otherwise erratic features of trends. In instances of parallel evolution the selectionist theory is that changes actually occurring in parallel are adaptive over the whole ecological range occupied by the group, while those divergent (radiating) within the group are adaptations to special niches within that range. At the least, that is plausible in such cases as that of the therapsids and earliest mammals. The parallel changes are involved in development of homeothermy, of more homeostatic and efficient metabolism, of improved sensory and central nervous functions, and of more efficient reproduction—all characteristics that can hardly have failed to be adaptive in those environments for any animals that were structurally and genetically able to develop them. The divergent characters are almost obviously specializations for particular and diverse ways of life.

The phenomena here discussed have been the principal evidence cited for vitalist, endogenous, or perfectionist theories of evolution. Those theories are basically appeals to ignorance or examples of the naming fallacy, because "entelechy," "inherent tendency," "aristogenesis," and the like have never been defined in terms of objective, material causes and, in fact, are usually advocated as names for unknown causes postulated as indefinable in such terms. We may, indeed, be ignorant of the causes of trends, or the causes may, indeed, be transcendental, but naming our ignorance does not alleviate it, and postulating the transcendental always stultifies inquiry. There is some material basis for judgment. If such views are true, there should be observable trend phenomena inexplicable in any other way. It has been claimed that the frequency or universality of indefinitely continued, invariant, nonadaptive trends is such a phenomenon; but we have seen that trends really evident in the fossil record do not have such characteristics and are readily explicable by known material causes.

As a biological term the word "evolution" was originally applied to ontogeny. It meant embryological development, especially but not exclusively under the theory of preformation. Only relatively recently, especially since Darwin, has "evolution" been understood as strictly phylogenetic. The terms "morphogeny" and "morphogenetic" still suffer from that ambiguity, being currently applied indifferently to the embryological origin of anatomical structure in individuals and to the phylogenetic development of structure in the course of evolution. Following the latter usage, paleontologists have been much concerned with abstracting from the fossil record so-called morphogenetic laws. These, far from being laws in the physical sense, are usually generalized descriptions of trends that have occurred widely and among numerous organisms only distantly related. In no case are these universal. They are in most instances merely frequent and not demonstrably usual.

Half a dozen generalizations of that sort will be considered in the course of Chapter 11. Just two of them will suffice as examples in the present different context. It is unlikely that others require different theoretical interpretations. The first is the tendency for the individuals in populations to become larger as evolution proceeds. This is probably the most widespread trend yet observed among animals, with known examples ranging from protozoans to man. Yet it is at least as common for organisms to persist without any consistent trend for change in size, and changes toward smaller size, although apparently unusual, have certainly occurred. There is, again, no real evidence in favor of mutational control or inherent tendency. In intraspecific competition under a great variety of circumstances, individuals slightly, but only slightly, above the population mean size are evidently favored by natural selection. Rensch has noted a variety of reasons for adaptive advantage in the trend toward larger size when it does occur.

The second example is of a kind of morphogeny less widespread but more complex and more interesting. This is the tendency for a series of similar anatomical parts, often segmental, to become reduced in number while the remaining parts become less similar (more differentiated in structure and function). Crustacean appendages are a striking instance, and there are many others. Here, too, exactly the opposite has also happened, although apparently much less frequently. For example, the vertebrae of snakes have become more numerous and less differentiated than were those of their legged ancestors.

Changes of those sorts illustrate particularly well a broad morphogenetic principle, as distinct from descriptive generalizations such as have now been exemplified. Historical change is always a modification of an existing configuration. In morphogeny it starts with a given adult structure, which is produced by a given developmental sequence, which, in turn, is controlled by a given genetic system. Mutations, the really new materials for evolution, affect the genetic end. Selection, the nonrandom element of evolution, acts anywhere or everywhere along the line. Both are circumscribed by what is already there, a configuration resulting from the whole past. The past is irrevocable, and therefore evolution is irrevocable, not only in the sense of irreversibility—an earlier configuration can never be regained—but also in that of indelibility—past configurations cannot be wholly effaced (except in a certain sense by extinction, but even extinct groups have influenced later configurations).

Nothing in the recorded history of life arises *de novo*. All is transformed from what went before. That is richly exemplified by transformation in a more technical sense: the adaptation to new uses of previous structures, physiological processes, and behaviors. Structural examples, the only ones *directly* visible in the fossil record, have innumerable examples: the famous transformations of gill arches to jaws and much later of certain bones of the lower jaw to auditory ossicles; the transformations of pectoral fins to forelegs and of forelegs divergently to flippers, wings, and arms—but why multiply examples when so many are now familiar and the list is almost endless. In physiology we need only mention the transformation of endostyle to thyroid gland or the changing targets of hormones. In behavior some of the ethological displacement activities are brilliant examples.

All science is philosophical, and the only philosophies capable of validation are those of scientists. Both scientists and philosophers frequently object to those statements, and some of their objections are valid, but the statements have a residual truth that cannot be wished away. A scientist cannot so much as make an observation without reliance on a philosophical premise, such as the by no means self-evident minimal premise that there really is something to observe! And, at the very least, his observations place restrictions on what can in any meaningful sense be true. What kind of universe is this we live in? It is, among other things, the kind in which life's history could and did occur. We can learn more about our kind of universe from that history than from pure reason and at least as much as (I think rather more than) from the stars or the atoms. Some of the evidence is now before you. In following chapters philosophical problems evoked by that history will be considered along with some further evidence. At this point I shall make rather flat statements of conclusions to be more adequately supported subsequently. The conclusions bear on four related problems: order, utility, progress, and purpose in the history of life.

1. The problem of order is that of uniformitarianism or of immanence in a special guise. The universe *is* orderly. It has certain built-in characteristics that came we know not whence or why but that are determinable and that have not changed during the course of recoverable history. This does not mean, as some uniformitarians have supposed, that configurations have only varied about a mean. There have been times when erosion of the earth's crust predominated and times when deposition was pre-eminent. The effects of those processes have been progressive, not merely fluctuating. Organic evolution has been faster and more basic at times in the past

than it is now or is ever likely to be again. But in both those and all other cases the immanent processes have been unchanging.

Of course, that cannot be *proved*. It fits the evidence, and it fits better than any opposing conclusion, but the evidence could itself be misleading. We cannot disprove the postulate that the universe was created one second ago, complete with all our apparent memories of our own earlier days, or that it was not created in 4004 B.C. with all the apparent record of earlier billions of years. But that would not make sense, and we must pretend, at least, that both we and the universe are sane.

2. The problem of utility is part of the problem of teleology, whether evolution has goals or ends and, if so, what and whose those ends may be. Again there seems to me to be only one answer, even though a somewhat complex one, that is thoroughly congruent with all the evidence and that is validated in that sense. The organization of organisms certainly has utility, and the evolution leading to them has that utility as a goal in a sense. That sense is, however, quite special and does not at all correspond with teleology in the classic meaning of correspondence with a preordained plan, with divine Providence, or with purposes especially relevant to the human species. The utility of any feature of organisms is with respect to the population (not invariably the individual) of those organisms at any given time. It is not related to usefulness to any other organisms; it follows no pre-existent plan; and it is not prospective toward any future goal. The over-all and universal goal is a posteriori at the given moment and is simply survival, which involves comparative success in reproduction.

This is not to say that all features of all organisms are useful to them at every moment in time. Some features reflect a utility now past, hence vestigial organs and many characteristics involved in transformations. The vast body of evidence for the irrevocability of evolution shows that the structures of organisms were not created or evolved, shiny and new, *for* the organisms but were evolved *by* the organisms from what history made available to them. Then, too, some features perhaps without utility have evidently been carried along by subtle connection with features that are useful. It is also possible, but debatable, that some characteristics have no connection with utility. It still is true (i.e., is the most reasonable conclusion from a vast body of evidence) that utility in the teleonomic, not teleological, sense here given the word is the principle that makes evolution orderly to the extent that it is in fact orderly.

3. The history of life is obviously, indeed tautologically, *progressive* by one definition. It proceeds by successive stages or, better, gradations through an indefinitely prolonged series of conditions (configurations), each derived from and differing from those preceding it, and that is a usual understanding of the words "progressive change." The problem is whether the history is also progressive in the quite different sense of involving, on the whole, change *for the better*. But "better" is an evaluation, meaningless unless one can designate better for whom and in whose judgment or in what sense. In line with our conclusion on utility, the most obvious reply would be, "Better for each separate population of organisms in the sense of being useful to it." But, a conclusion on utility having been reached, that further statement is banal and adds nothing of interest. It does not begin to tell us what we really want to know.

If a useful function comes to be performed more effectively or if a new function adds to utility, then there has been improvement. It is meaningful to consider such improvement as progress, and, while involving the concept of usefulness, that statement does add something to the previous conclusion on utility. The limb of a mammal cannot be considered an improvement over the fin of a fish, because neither performs more effectively the functions of the other or adds (without equivalent loss) to the functions of the other. But some fins function better as fins than others, and some legs as legs. There is improvement among fins and among legs, but not between the two. What is better depends on where they are. Some improvements are far more general: the vertebrate eye is in an extremely broad sense an improvement over a protozoan's light-sensitive pigment spot. That is still not a completely general example. The vertebrate eye is an improvement only if accompanied by a nervous system able to organize the sensations received and a behavioral system able to make that information practically useful.

Hence one improvement requires others, or it is not, in fact, an improvement and will not arise in the course of evolution. Plants have their own improvements, and it would be no improvement for them to have eyes even if a vertebrate-like nervous system accompanied them. The only universal improvements are in those features common to all organisms and involved in the origin of life.

With those definitions and restrictions, progress is certainly a usual feature of the history of life. Most of the events of evolution seem to be either improvements or transformations, using "transformations" here more in the sense of breakthroughs which involve changed functions and hence changed avenues of improvement. This is not contradicted by the fact that progress can be absent or reversed in certain instances, and it is only superficially paradoxical that the absence or reversal may even be useful.

Up to now, at least, man stands as the high point of evolutionary progress. He can do many more things than any plant and can do almost anything that any animal can do, and generally better. The ability to use tools and artifacts in this connection is, of course, one of man's *biological* improvements.

4. There is no fact in the history of life that requires a postulate of purpose external to the organisms themselves. It could, of course, be maintained that the whole system, purposeless itself, was created for a purpose or that purposes not required by the evidence may nevertheless exist. Such speculation is without control, incapable of validation, and therefore altogether vain. We do know, however, that purposes peculiar to and arising within organisms exist as one of the great marvels of life. We know it because we form purposes ourselves. We do not know how general such purposes are among organisms. Must a purpose be conscious, and, if so, how far does consciousness extend? Perhaps it is only a matter of definition, and perhaps there is some sense in which purpose is one of the universal improvements of living over nonliving.

CHAPTER NINE

Evolutionary Determinism

THE rise of modern science was characterized by the formulation of laws that were, or seemed to be, absolutely deterministic. In the physical sciences, experimentation showed that defined circumstances produced predictable results with a certainty well within limits of observational error. These laws, such as the law of gravity or the law of combining weights, could be expressed mathematically. Their operation could be demonstrated by anyone and could be repeated as often as was desired. The variables involved were typically few, simple, and easily isolated. The passage of time did not prevent repetition of all the pertinent causal factors and produced no perturbation in the results. Gravity has no history.

The predictive value of such laws seemed to be 100 per cent. Although, in fact, the established laws of this sort covered only a small fraction of the phenomena of the universe, it was easy to conclude that the progress of research would discover other laws until the sway of mechanistic determinism would be found complete. Many agreed with Laplace that if we knew all about the universe at any given moment, we could predict exactly what would happen forever after.

Everyone knows that the physical scientists are no longer convinced of the universality of such absolute determinism. Indeterminacy has crept in on two different levels, which do not always seem to be clearly distinguished. Experimental verification and prediction may be impossible because the pertinent variables cannot be simultaneously observed. Apparent determinacy of events may be merely the extremely probable but not absolutely certain outcome of statistical effects, averaging very large numbers of individual cases which are not separately determined by the same laws. (That the individual events are therefore not strictly determinate by any laws whatever seems to be a conclusion of faith rather than of logic or evidence, but that is beside the present point.) This hedging on determinacy has considerable importance when delving into the atom or into the far reaches of the cosmos. It has also changed attitudes as to the basis for physical laws, but it has not really abrogated those laws. The gas laws may be merely statistical, but they continue to apply with full rigor to practicable observations in the laboratory.

The idea of a completely lawful universe was part of the intellectual atmosphere in which the theory of evolution was born and grew up. Evolution was commonly hoped or believed to be a process as orderly as, say, a chemical reaction. It was easy to assume that there must be a set of laws of evolution, absolute, universal, and deterministic, the discovery of which would provide a complete explanation of the history and present condition of life. If such laws have determined the history of life, they should be discoverable by examination of that history and especially of the fossil record. In this field, at least, the question thus becomes definitely one for scientific investigation and not one for philosophical speculation only. A considerable number of supposed laws of evolution were, indeed, promulgated by paleontologists and others; several of these "laws" will be named and considered later.

Some students would be willing to go all the way with Laplace and to hold that the laws of evolution are predictive, that the future history of life is (theoretically!) wholly implicit in the present moment. Probably most paleontologists, inclined to retrospection more than to prediction, would prefer to define absolute determinism as the belief that everything that has happened in the history of life had to happen; it could not have been otherwise. Evolution might then be likened to the working out of an equation with a unique solution.

Even within the fold of what can properly be called determinism, this absolute, Laplacian determinism does not exhaust the possibilities. Evolution might be *determined*—that is, completely caused in a materialistic way—and yet not rigidly *predetermined* from the first as to the course it was to follow. An equation can have multiple solutions, and yet each solution is determined by the equation. Then any laws, or rather determining principles, of evolution would fall short of full predictive value. The possibility of a nonpredictive determinism is enhanced by the fact that prediction in the physical sciences depends on the unrestricted repetition of causes. If causes cannot be precisely reproduced, then results cannot be precisely predicted.

Then, too, there is the possibility that evolution is partly deterministic, partly not. It may combine physicochemical determinacy with biological indeterminacy. This is the view of some vitalists, such as Vandel, for whom evolutionary determinism is a *loi approchée*. (In some current scientific circles it is considered old-fashioned or not quite nice to be a vitalist, so Vandel and some other vitalists do not accept this label on the package, but they supply pure vitalistic contents.)

The vitalists want to live in a dualistic universe with two essentially different sets of forces, principles, and essences: matter and the material forces, life and the vital forces. The finalists, who are usually also vitalists, want further to live in a purposeful universe. They want evolution to have had a goal, and they think that the history of life has been a means to an end, an effect preceding its cause as far as material manifestations occur. Both vitalists and finalists tread close on the heels of theology, for "life essence" may be soul, and the "purpose" in evolution may be God's.

In their bearing on determinism, vitalistic and finalistic views have involved a strange and almost surreptitious reversal of roles. Historically, materialism, mechanism, and determinism have been closely associated. To this day the three terms are often virtually synonymous in popular estimation and are likely to be considered as essentially associated by many scientists and philosophers. The curious reversal is that in evolutionary theory, the materialists have often tended toward indeterminism and the nonmaterialists (mostly vitalists, finalists, or both) toward determinism. These terms are not used and the issues have not been clear, but the confusion is largely semantic and the arguments are often mere obfuscations.

There have been two main currents of more or less materialistic evolutionary theory. One has run from Buffon, Lamarck, and St. Hilaire through Neo-Lamarckism to Michurinism, further notice of which is deferred for a moment. The other has progressed from Darwinism through Neo-Darwinism and mutationism to the modern synthesis (also often confusingly labeled Neo-Darwinism). Proponents of these latter schools have made much of the partly or wholly random nature of what they take to be the materials of evolution: hereditary variation and mutation. Their opponents have accused them of believing in evolution by pure chance. This is not valid criticism because only a few extreme mutationists have considered evolution to be wholly random, but the point is that both the adherents and the critics of this current of evolutionary thought have emphasized its inclusion of random or chance elements in evolution. Although this need not rule out determinism of some sort or degree, it is hardly consistent with absolute, mechanistic determinism.

On the other hand, the vitalists, and especially the finalists (often but not always the same persons as the vitalists), have developed under other names a rigid determinism of their own. This is evident from the very fact that their basic criticism of materialistic theories is that the latter rely on chance altogether (which is not really true of most materialistic theories), or at least more than the evidence warrants. They speak of "antichance" as a peculiarly vitalistic or finalistic element in evolution, and it may be significant that the term is consciously borrowed from a physicist (Eddington). Cuénot, who among finalists is one of the few with a genuine, broad grasp of really pertinent evidence, recognizes both chance and antichance in evolution. He assigns chance to the mechanistic aspects of life, antichance to the finalistic, an unusually clear demonstration of the reversal of the usual associations.

All adherents of vitalist-finalist views adduce as primary evidence the supposed phenomena of orthogenesis, and define orthogenesis as a force (never really explained) impelling evolution to proceed only along fixed, predetermined lines. Osborn, another learned although somewhat obscurantist vitalist and finalist who certainly had tremendous knowledge of the evidence, held that *"fortuity is wanting"* in evolution. In this he spoke for many of the nonmaterialists. Surely a process without fortuity is deterministic, and it becomes clear that theirs is a theory of vitalistic determinism no less rigid than mechanistic determinism.

The role of Michurinism is also peculiar and should be briefly mentioned. It is ideologically improper for Soviet biological science to be other than materialistic and Darwinian. But Darwin and most of his followers have not been wholly and rigidly mechanistically deterministic. Somehow materialism and determinism seem to be related concepts. From this sequence of conceptual associations, a dilemma arises for those who hold that science must be controlled by political Soviet ideology. The 1948 session of the Lenin Academy established Lysenko as high priest and political boss of Soviet biology and Michurinism as its gospel. One of Lysenko's clinching arguments was that the Neo-Darwinians see nature as a "chaos of chance" and that Michurinism banishes chance. In other words, Michurinism is claimed to be deterministic, and this is promoted as an alternative to the indeterminacy of modern geneticists and others (mainly distinguished Soviet scientists who were genuinely materialistic and Darwinian).

This attempted escape from the dilemma involved in current concepts of determinism and materialism has plunged Lysenko and the Michurinists into a worse dilemma still, and even after fifteen years it remains to be seen how soon this will become apparent to the Party bosses and what will be done about it. Michurinism is really vitalistic and not fully materialistic—on this we have, among other authorities, the interesting agreement of George Bernard Shaw, who approves of vitalism, and of H. J. Muller, who does not.

Again the odd linking of vitalism and determinism appears. Moreover, Michurinism is a profoundly reactionary doctrine. It is largely pre-Darwinian and almost entirely non-Darwinian. Calling it "Soviet creative Darwinism" is such obvious double talk that it can hardly obscure the facts indefinitely. That Michurinism is reactionary, non-materialistic, and non-Darwinian has no conclusive bearing on whether it is true, but at least the strictly scientific issues are clarified by placing it in its real position among biological theories.

Let us now turn more directly to the evidence of the fossil record on the main problem. Does this record suggest a deterministic control of evolution and, if so, in what way and to what degree? The pertinent data consist largely of regularities constantly or repeatedly appearing in the record. Since these regularities, real or imagined, are summarized in generalizations read from (or into) the record by its students, our aim here may be brief examination of the validity and generality of some of the proposed laws and supposed tendencies.

It is hard to pin down the more rigid formulas of evolutionary determinism and to put them in concrete and testable form. Thus Berg, author of *Nomogenesis; or Evolution Determined by Law* and among the most absolute of evolutionary determinists, has little to offer except the law that evolution is determined by law. The evidence put forward bears mainly on the claim that evolution has followed determined lines, that it has been orthogenetic. No satisfactory suggestions are made as to the nature of the determination, why one line was followed rather than another, or what the determinative instrumentality may be. Such vagueness or evasion is usual among the vitalistic and finalistic determinists, few of whom are so frank as Osborn, who maintained that the causes of evolution are quite unknown, although, in his opinion, the course is orthogenetic and directed toward a goal ("definite in the direction of future adaptation").

The crucial point here is whether evolution is in fact orthogenetic, whether orthogenesis is its law. There are many definitions of orthogenesis, but the pertinent one in this connection is that orthogenesis designates not only full determination of evolution but also rigid and unique predetermination. Orthogenetic evolution is supposed to proceed undeviatingly in a single direction, regardless of environment, organic activity, or such factors as natural selection. Discussion of this point has been so lengthy and extensive that it has, frankly, become boring. There is at present a clear consensus of paleontologists that orthogenesis, in this sense, is not real. There is no known sequence in the fossil record that requires or substantiates such a process. Many examples commonly cited, such as the evolution of the horse family or of sabertooth "tigers," can be readily shown to have been unintentionally falsified and not to be really orthogenetic. All supposed examples are more simply and fully interpreted as due to some other cause, such as natural selection. The fossil record is now usually cited in support of orthogenesis mainly by those least familiar with that record.

The fossil record definitely does not accord with the particular sort of rigid determinism that has come to be associated with the concept of orthogenesis or more broadly with overtly or covertly nonmaterialistic theories like those of Driesch, Bergson, Osborn, Cuénot, Lecomte du Noüy, or Vandel. There remains the possibility of materialistic determinism, which is a more open question.

A large number of evolutionary laws or principles have been proposed outside the field of orthogenesis and vitalism. As far as I know, no one has ever attempted to compile them all, and few have attempted to select among them a consistent set that might be supposed to encompass the determination of evolution as a whole. Perhaps the most logical compilation is that by Sewertzoff, who, with Vavilov, Dubinin, Schmalhausen, and others, is among the progressive Russian students of evolution whose work is anathema to the reactionaries now in control of the Party line in biology. His outline is brief and is worth reproducing (Table 1).

This list of principles governing functional changes associated with morphological changes in evolution does not, in fact, reflect any regularity or necessity. It is a classification rather than a code, and its statements are descriptive rather than prescriptive. As far as there is any bearing on determinism, the list might suggest some degree of indeterminacy in evolution. All of Sewertzoff's "types"

Table 1

Types or Principles

I.

The ancestral function of the evolving organ remains qualitatively the same in the descendants, but becomes intensified.

1. Intensification.
2. Substitution of organs, of N. Kleinenberg.
3. Physiological substitution, of D. M. Fedotow.
4. Fixation of phases.
5. Decrease in number of functions.

II.

The ancestral function of the organ becomes qualitatively changed in the descendants.

1. Broadening of functions, of L. Plate.
2. Change of function, of A. Dohrn.
3. Assimilation of functions.
4. Activation of functions.
5. Immobilization.
6. Substitution of function.
7. Division of organs or functions.

seem to be real phenomena and all can be illustrated from the fossil record, but they include alternatives or indeed tend to embrace all the possibilities. "Broadening of functions" is (with "no change" as an understood third possibility) the only alternative to "decrease in number of functions," "immobilization" is the opposite of "activation," and so on.

It is a "law" that, say, an evolving group of animals will increase in individual size or decrease or remain the same size, but this does not help to establish how or whether the outcome is determined. Such is the nature of Sewertzoff's principles and similar classifications of evolutionary phenomena. They are multilateral or all-embracing. Restrictive principles or laws are more likely to bear on problems of determination. As a matter of fact, one of the proposed restrictive "laws" of evolution is that evolving animals become larger. This sort of "law" is related to regularities in the fossil record, and it implies determined limitation of possibilities. A few of the more interesting of the supposed laws or principles of this limiting sort may now be considered:

1. Primitive forms survive longer than specialized forms (sometimes called "Cope's law of the survival of the unspecialized").

2. Body size increases in the course of evolution. (This has also been called "Cope's law," as well as "Depéret's law.")

3. At a given time, different evolving groups change in the same way (Dacqué's principle of *"Zeitsignaturen"*).

4. In the course of evolution, the number of structural parts becomes reduced and the remaining parts become more specialized (usually called "Williston's law," but Stromer thinks it should be "Stromer's law").

5. Evolution is cyclic, and evolving groups go through phases of youth, maturity, and old age (a "law" variously expressed by Haeckel, Schindewolf, and others).

6. Evolution is irreversible ("Dollo's law," but as in most cases the principle was known before its notice by the student for whom it is named).

Although these "laws" are here given in categorical form, most of their authors and discussants have qualified them, and qualification is necessary.

1. The survival of the unspecialized is certainly frequent in the fossil record. The opossum, little changed since the Cretaceous, has outlived many more specialized relatives. Numerous other examples are known. On the other hand, extinction of the less and survival of the more specialized seem to be about equally common. The living elephants, for example, are more specialized than the extinct mastodons, and in general all surviving ungulates are more specialized than many extinct allies.

2. Increase in body size is very common, a stock example being the change from eohippus to the modern horse. The phenomenon is perhaps sufficiently usual to be a rule, but the rule has many exceptions. Even in the horse family, several evolving lines became

smaller rather than larger. The apparent extent of the rule has been exaggerated by students who thought it absolute and who insisted that because an earlier animal was larger than a later relative *therefore* it was not ancestral to the latter.

3. It has never been claimed that "program evolution," or *Zeitsignaturen,* tendencies for different lines to evolve in the same way at the same time, is universal. This has occurred, usually in the form of parallel evolution in more or less closely related lines, but as a matter of fact it is not so usual a phenomenon of evolution as to be considered a rule or the usual tendency.

4. "Williston's law" is well known to be an occasional tendency and not a universal principle. Its application is more limited than usually stated, because it works out only when an ancestral form has multiple parts performing essentially the same function, as in the dermal bones of the vertebrate skull or the legs of crustaceans. Under these particular conditions, reduction in number and increase in specialization of the remaining parts are clearly common and probably are usual. It is, however, well known that the opposite, multiplication of parts, is also a common evolutionary trend.

5. The idea that evolving groups have a phylogenetic life cycle has frequently been expressed, and numerous students have given names to phases of the supposed cycle. Three phases are usually recognized, but the names and descriptions of these phases vary widely among authors. A recent example is Schindewolf's distinction of "typogenesis," "typostasis," and "typolysis" as the three phases of (in his opinion) a universal and cyclic sequence in phylogeny. Phylogeny is certainly episodic, with evolution moving faster at some times than at others and with occasional periods of relatively rapid divergence ("explosive" evolution or adaptive radiation), commonly followed by slower progressive specialization of the various lines and by extinction of some of them. I can, however, find no good evidence that the episodic nature of evolution is truly cyclic in any regular way or that it has phases that can be called young, mature, and old in any reasonable sense of the words. The episodes of "youthful" divergence (typogenesis of Schindewolf) occur most irregularly and may recur within the history of a single group with-

out intervening "old" stages. In the horse family, for instance, there have been at least three major episodes of this sort. The fossil record as I read it also lacks any valid evidence that lines of evolution reach old age in the sense of inherent waning of vital potential or that the claimed phenomena of "gerontism" are in any true sense indicative of terminal phases of a phylogenetic life cycle.

6. The principle of irreversibility in evolution is on a different footing from the others here listed. It is more general and more important for our subject. The principle frequently is not true of broad trends or of changes in particular structures and characters. A trend toward larger size, for instance, is readily reversible. A specialization, such as enlargement of a tooth or development of hoofs from claws, may also be reversed. A whole group may converge toward a remotely ancestral group, as whales toward fishes, and there will then be some reversion in adaptive characters. In these respects and within limitations as to exactness of duplication, evolution is reversible and the principle has misled students who thought that it applied to such cases. In a broader sense, however, the principle is true statistically if not absolutely. There are no known cases in which a structure, of any noteworthy complexity, at least, has reverted *precisely* to a distinctly different ancestral condition. With less qualification, there are no known cases in which a type of organism that had become extinct or considerably modified was ever reduplicated.

That evolution is irreversible is a special case of the fact that history does not repeat itself. The fossil record and the evolutionary sequences that it illustrates are historical in nature, and history is inherently irreversible. Realization of this fact puts the question of evolutionary laws and of determinism on quite a different basis.

Physical or mechanistic laws depend on the existence of an immediate set of conditions, usually in rather simple combinations, which can be repeated at will and which are adequate in themselves to determine a response or result. In any truly historical process, the determining conditions are far from simple and are not immediate or repetitive. Historical cause embraces the *totality* of preceding events. Such a cause can never be repeated, and it changes from instant to instant. Repetition of some factors still would not be a repetition of historical causation. The mere fact that similar conditions had occurred twice and not once would make an essential difference, and the materials or reagents (such as the sorts of existing organisms in the evolutionary sequence) would be sure to be different in some respect.

It is impossible to observe whether Cambrian trilobites would have evolved in the same way under the same conditions in the Ordovician. Cambrian trilobites did not exist in the Ordovician and could never again exist after the Cambrian. Nor were the conditions in which they evolved present in the Ordovician or at any time after the Cambrian.

Recurrence of more or less similar conditions within the great complexity of historical causation may evoke some similarities of response, but these will be flexible, nonidentical, and representative of tendencies rather than of invariable responses to law. "Laws" like those cited above clearly generalize such tendencies and are not laws in the usual mechanistic sense. At most they are descriptions of what has commonly happened, or in some cases only of what has sometimes happened. The fossil record is consistent with historical causation that is in continuous flux, nonrepetitive, and therefore essentially nonpredictive. It is not consistent with absolute, repetitive, and predictive determinism.

History does not correspond with possible mechanistic models such as serve (with certain reservations, of course) in the physical sciences. That history is not simple and tidy is unfortunate, perhaps, but it is true. Probably all teachers of historical subjects have had students whose training or bent is in the physical sciences and who have become bewildered or antagonistic on learning that the important questions of history cannot be answered by an equation or an experiment. The human desire for neat and unequivocal conclusions explains the long and necessarily futile search for simple, absolute, deterministic laws of evolution.

The record does show beyond any doubt that there are directional forces in evolution. There are trends and there are common tendencies. These are reflected in the "laws" that have been proclaimed from time to time, but those "laws" are not laws. The regulation, which seems surely to be a determination, is by forces that change continuously in intensity, direction, and combination, and that produce quite different results in different instances. These forces are themselves interwoven with the historical process and subject to historical causation, rather than being absolute or unchanging, in the sense of physicochemical laws, or yet internally inherent (as in vitalism) or metaphysically predetermined (as in finalism). Only two possible sorts of forces are consistent with these requirements of the record. One sort involves the Neo-Lamarckian factor, the inheritance of changes directly induced by organism-environment reaction. This possibility has been thoroughly tested both by consideration of the fossil record and by experimentation, and it has necessarily been discarded, except in the untenable and retrogressive doctrine of Michurinism. It is inadequate to explain the regulation apparent in the fossil record, and it is experimentally disproved.

The other possibility is natural selection, and this must be accepted, not by elimination, but because it is entirely consistent with the fossil record, adequate to explain the regulation evident in that record, and experimentally verified. Like the Neo-Lamarckian factor, formerly postulated but now known to be nonexistent, natural selection arises in the material historical process. It is also an interaction between organism and environment the results of which, in itself, are determinate, although the process of determination is extremely complex, far more so than if the Neo-Lamarckian factor were really operative.

Beyond this point, natural selection is characterized by not producing its results directly. It can only act on materials presented for its action, which are genetic variations. Given an existing population structure and an existing ecological situation, and given the genetic variation of the population as it moves through time, the action of selection seems to be fully deterministic. The first two elements, population structure and ecological situation, are themselves entirely *historically* determined. Genetic variation apparently is not fully determined in the same way, or, rather, if it is so determined, we do not know how. As now known, both mutations and the genetic effects of sexual reproduction have a random element, in the sense that they have aspects for which historical causation cannot be assigned. These aspects do not seem to be consistently oriented with respect to the direction of natural selection. There is no good reason to doubt that genetic variations are materialistically caused, although rigorous proof is lacking. Yet the fact that they are in part random in relationship to the historical determinant of natural selection means that they make possible multiple solutions of evolutionary problems.

All this fits in perfectly with the known fossil record. Trends exist, but they are flexible and are not precisely repeated. Different results do follow similar causes or appear as solutions of the same adaptive problems. There are limitations, but they are not absolute. There is direction, but it wavers, and apparently random effects also occur.

(One more suggestion is thrown in, tentatively and parenthetically. As it affects some organisms, especially higher animals and most particularly man, historical causation may involve anticipation. Not only the totality of the past but also, at these levels, something of the future may thus be involved in evolutionary determinism. I do not, of course, refer to the metaphysical finalistic idea of a future goal which determines preceding events, but to the material fact that man and some animals sometimes act in anticipation of events that have not yet occurred. The anticipated events may or may not occur when the time comes. The possibility of multiple solutions and of nonpredictive determinism may thereby be enhanced. Anticipation is completely absent in physicochemical causation, and it adds to the reasons why evolutionary "laws" are not like physicochemical laws. Anticipation is nevertheless a fully materialistic process.)

Evolution is in part demonstrably deterministic. It is presumptively so in its entirety, but only in a special sense of determinism. The peculiarity of evolutionary determinism consists in its being historical and not mechanistic and in its permitting multiple solutions and not only a unique outcome. It is therefore both nonrepetitive and nonpredictive.

CHAPTER TEN

Plan and Purpose in Nature

"OF surpassing interest to those many minds, which seek after philosophic knowledge and instruction, is the Story of the Earth, Her manifold living creatures, the human generations, and Her ancient rocks." With these words Doughty introduced the second edition of his great book *Travels in Arabia Deserta*. They might serve as a preface to all human learning and especially to that most important of the branches of learning, the study of life. "The human generations," in all their aspects, must be the constant concern of any thoughtful man, whether he aims at "philosophic knowledge" only to enrich his own life or whether he aims at "instruction" to guide him in his functions as a member of society.

In degree, man's social and intellectual complexity is something new under the sun, but man remains a part of nature and is subject still to all of nature's laws. Man is only one of earth's "manifold living creatures," and he cannot understand his own nature or seek wisely to guide his destiny without taking account of the whole pattern of life.

We feel, almost instinctively, that there is a pattern. The diversity of living creatures is neither complete nor random. All living things share many characteristics, and above this basic level we observe groups with every degree of resemblance, from near identity to great dissimilarity. There is, or seems to be, an essential order or plan among the forms of life in spite of their great multiplicity. There seems, moreover, to be purpose in this plan. The resemblances and differences among a fish, a bird, and a man are meaningful. The resemblances adapt them to those conditions and functions that all have in common and the differences to peculiarities in their ways of life not shared with the others. It is a habit of speech and thought to say that fishes have gills in order to breathe water, that birds have wings in order to fly, and that men have brains in order to think.

A telescope, a telephone, or a typewriter is a complex mechanism serving a particular function. Obviously, its manufacturer had a purpose in mind, and the machine was designed and built in order to serve that purpose. An eye, an ear, or a hand is also a complex mechanism serving a particular function. It, too, looks as if it had been made for a purpose. This appearance of purposefulness is pervading in nature, in the general structure of animals and plants, in the mechanisms of their various organs, and in the give and take of their relationships with each other. Accounting for this apparent purposefulness is a basic problem for any system of philosophy or of science.

Attempts to solve this problem are perhaps as old as man. There are few savages so dull or so primitive that they do not have some legend or belief bearing on the problem, even though they do not formulate it clearly or in these terms and even though their solutions are more implicit than explicit. Certainly the problem appears repeatedly, in more or less sophisticated form, among the Greek philosophies, in the scriptures of all the great religions, and in other ancient attempts to grapple with the nature of the universe and of man. In scientific history, the need for a solution became particularly clear and conscious in the intellectual crisis of the latter part of the eighteenth and early part of the nineteenth centuries.

One possible solution, which had by then become traditional in Western thought and religion, was elaborately presented in the *Bridgewater Treatises*, published in 1833–40. These treatises, of which eight were issued, were richly endowed under the will of the eighth Earl of Bridgewater, who directed that they set forth

> . . . the Power, Wisdom, and Goodness of God as manifested in the Creation; illustrating such work by all reasonable arguments, as for instance the variety and formation of God's creatures . . . the construction of the hand of man, and an infinite variety of other arguments. . . .

They were of unequal merit, but several of them can still be read with pleasure and, indeed, profit. The best of them were, in their day, able works of science, and they strongly influenced scientific thought. (It is noteworthy that although the *Bridgewater Treatises* were, by stipulation, antievolutionary, a quotation from one of them stands at the beginning of Darwin's *The Origin of Species*.)

The work on the hand, particularly mentioned in Bridgewater's instructions, was written by Sir Charles Bell under the title "The hand, its mechanism and vital endowments as evincing design." Bell did not confine himself to the human hand, but gave a generally excellent account of the comparative anatomy of the vertebrate forefoot and, indeed, of handlike appendages throughout the animal kingdom. He stressed the perfection with which each type of forefoot is adapted to the particular needs and habits of its owner and he pointed out that the intricate mechanism of the human hand follows a seemingly perfect and (as it appeared) obviously purposeful design.

> There is an adaptation, an established and universal relation between the instincts, organization, and instruments of animals on the one hand, and the element in which they are to live, the position which they hold, and their means of obtaining food on the other.

After considering and rejecting the idea of evolution, which was, of course, already known at that pre-Darwinian time, Bell concluded:

> It must now be apparent that nothing less than the Power, which originally created, is equal to the effecting of those changes on animals, which are to adapt them to their conditions: that their organization is predetermined, and not consequent on the condition of the earth or the surrounding elements.

The fact of adaptation was thoroughly established by such works as Bell's. The tremendous increase in knowledge of nature since 1833 has, on this point, served only to demonstrate that adaptation is even more widespread and may be even more elaborate than Bell knew. To this extent, his arguments are just as cogent now as when he wrote them. But now that we know that evolution is a fact, we can no longer accept his simple solution of the problem of adaptation as reflecting the purpose of a Creator manifested in the separate creation of each species of animal or plant. Whether or not we can explain the evolution of adaptation has no necessary bearing on the truth of evolution. The proofs that have now accumulated, quite aside from attempted explanations of adaptation, are fully sufficient. Competent modern biologists may differ as to the meaning or mechanism of adaptations and yet all agree that these did, somehow, arise by evolution. If, however, we hope—as of course we do—to go beyond merely observing the course of evolution and to gain some insight into how and why nature has followed this course, then we must still account for the appearance of purposefulness in this history.

In Bell's day, before the proofs of evolution had accumulated sufficiently to carry full conviction in themselves, his argument could only be met by demonstrating the possibility, at least, of an evolutionary explanation of adaptation. Even now we are, quite properly, I believe, reluctant to accept the reality of a phenomenon, be it evolution or extrasensory perception, unless there is some plausible hypothesis as to its mechanism. It is, then, not surprising to find that both Lamarck and Darwin, whose ideas in more or less profoundly modified form have dominated most of the subsequent discussion of evolutionary theory, were at least as much concerned with finding an evolutionary mechanism for adaptation as they were with demonstrating the truth of evolution.

Lamarck's theory was discussed at length in Chapter 3. It was there noted that Lamarck considered specific adaptation as something superposed on and divergent from the main course of evolution. Nevertheless it is this part of his theory that has influenced later thought on adaptation and that is pertinent to the present chapter. In this context, the Lamarckian theory may be expressed in five propositions, given as follows in his own words but translated and rearranged:

1. Nature, in successively producing all the species of animals, commencing with the most imperfect or the most simple and ending her work with the most perfect, has gradually complicated their organization.

2. If the cause that tends constantly to complicate organization were the only one that influenced the forms and organs of animals, the increasing complication would be, in its sequence, very regular throughout. But this is not the case. Nature is forced to submit her works to the influences of the environment, which acts on them, and on all sides the environment causes variations in the products of nature.

3. Whatever the environment may be, it does not directly bring about any modification whatsoever in the form or organization of animals. But great changes in the environment lead to great changes in the needs of animals, and such changes in their needs necessarily lead to changes in their actions. Now, if the new needs become constant or very long continued, the animals acquire new habits, which are as lasting as the needs that gave rise to them.

4. *First law:* In any animal that has not passed the limit of its evolution, the more frequent and continued use of any organ strengthens this organ gradually, develops it, enlarges it, and gives it a power proportionate to the duration of this use, whereas the constant lack of use of such an organ gradually weakens it, deteriorates it, progressively diminishes its capacities, and finally makes it disappear.

5. *Second law:* Everything that nature has caused individuals to gain or lose through the influence of the environment to which their race has long been exposed and, consequently, through the influence of the predominant use of an organ or that of constant lack of use of a part, nature keeps all this by heredity in the new individuals that are born, provided that the acquired modifications are common to the two sexes, or to the progenitors of the new individuals.

Most Neo-Lamarckians have rejected or overlooked the broader part of Lamarck's own theory and have supported a general theory of evolution based on the two "laws" that, by Lamarck, were meant only to explain exceptions to the normal course of evolution. Many Neo-Lamarckians have also postulated direct effects of the environment on the organism, which were categorically denied by Lamarck. Although I have summarized Lamarck's whole theory here, in an effort to restore historical accuracy, it is mainly his first and second "laws" that are now of interest to us, also.

Without using that term, Lamarck accepted the purposefulness of adaptation, but he reduced this to evolutionary and (at least superficially) mechanistic terms by equating the purpose with the needs of individual animals and their efforts to supply these needs. The giraffe has become a stock example, so used by Lamarck himself, and later by Darwin for his different explanation. The "purpose" of a giraffe's long neck is to reach higher leaves and twigs on trees for food. According to Lamarck, the ancestral giraffes stretched their necks in efforts to reach this food—purposeful behavior by the animals. In time, by long repetition, this elongated their necks, the elongation was passed on to their offspring, and this process continued for many generations until it culminated in the long necks of the present giraffes.

This theory of adaptation is ingenious and elegant. Its very simplicity has great appeal, and there is in it something of aesthetic satisfaction, a sort of poetry of evolution. Although it had little influence when first promulgated by Lamarck, later, when evolution had become generally accepted, it was widely discussed and commonly espoused, although it probably never represented a majority view on adaptation. It had a particular attraction for the paleontologists of the latter half of the nineteenth century. They found that adaptation was a key to many of their observations and that it was as widespread in the past as in the present. They also found many examples of the gradual, progressive perfection of adaptations that seemed beautifully explicable in Lamarckian terms, while they had difficulty in explaining some of these phenomena by the Darwinism of their days.

It is a pity that this thoroughly charming theory is not true, but there can now be no serious doubt as to its falsity. The theory indispensably demands that the effects of use and disuse, or other acquired characters, be inherited and that they be inherited in kind and become germinally fixed. That is, Lamarckism requires not merely that modifications acquired during the lifetime of parents should affect their offspring but also that they should affect the same parts of the offspring in the same way as in the parents and that they should, sooner or later, become a permanent part of heredity in the line of descent, regardless of further or repeated modifying factors. It has been proved, as nearly as a negative can ever be proved, that this does not and cannot occur. Even aside from this fatal point, the evidence against Lamarckism is convincing. There are adaptations, such as those of neuter insects and many cases of protective coloration or mimicry, for which a Lamarckian explanation is practically inconceivable. The adaptations, like the giraffe's neck, for which Lamarckism does ostensibly provide a mechanism are readily explicable by other processes which, unlike Lamarckism, have been experimentally verified.

I have digressed briefly to show that we cannot now adopt the Lamarckian explanation of apparent purpose in nature, much as we might like to do so. Historically, the greatest vogue of Lamarckism was in the future when Darwin advanced an alternative theory. *The Origin of Species by Means of Natural Selection; or, the Preservation of Favored Races in the Struggle for Life,* to give Darwin's greatest work its full Victorian title, was published in 1859, just a half-century after Lamarck's *Philosophie Zoologique.*

It has been pointed out in previous chapters that Darwin's book accomplished a number of different things. Here our main concern is that *The Origin of Species* advanced a theory to account for adaptation in nature, and particularly progressive adaptation. This was the main subject of the book, and Darwin considered it his major contribution, although he recognized that even this was not completely original. The theory is, of course, that of natural selection. The organisms likely to have more descendants are those whose variations are most advantageous as adaptations to their way of life and to their particular environment. Thus evolution is likely to move in the direction of greater or more nearly perfect adaptation, and thus the fact of adaptation, purposeful in aspect but impersonally mechanistic in origin, is explained.

Darwin did not believe that natural selection is a complete explanation of evolution, or even of adaptation. Some students have made much of the fact that he did not exclude the possibility of the inheritance of the effects of use and disuse in Lamarckian fashion. This point should not, however, be much emphasized. Darwin was wise enough to see that parts of the evolutionary process were still mysterious, and cautious enough not to rule out completely any supplementary factor that could not then be disproved. He ascribed only an indefinite and very subsidiary role to the effects of use and disuse and he showed that some supposed examples (like that of the giraffe's neck) could better be explained by natural selection.

As Darwin foresaw, his work was followed by the rejection of special creation as a scientific theory. That special creation remains, in some circles, a theological tenet is beside the point here. It is also beside the point, but should perhaps be mentioned in passing, that later scriptural exegesis found no difficulty in accepting evolution, rather than special creation, as God's method of creation and that there need be no conflict between rational evolutionists and rational theologians. After *The Origin of Species,* independent proofs of evolution piled up at an increasing rate, and, equally important, Darwin had demonstrated, more successfully than Lamarck, the possibility of an evolutionary answer to the theological argument of purpose in nature.

The Darwinian theory of adaptation was, however, soon under fire from other convinced evolutionists. One of their main lines of attack was that natural selection, as advanced by Darwin and his successors, is a purely negative process. It destroys but does not create. It eliminates disadvantageous variations but tells us nothing about the origin of advantageous variations and therefore does not, after all, explain how adaptation arises. This merely indicated that natural selection is not the whole story of evolution, which Darwin never claimed it was. It might not have been a very important criticism of Darwinism as Darwin understood it, but some of the Neo-Darwinians, toward the turn of the century, did so overemphasize a narrow concept of natural selection that this objection did tend to discredit the whole Darwinian school. Aside from this development, the criticism is crucial to our present inquiry because the problem of purpose does not arise from the elimination of the unfit but from the origin of the fit, and it was claimed that the Darwinian explanation of the latter point was discredited.

It was also maintained that characters destined to be adaptive are often of no real use when they first appear, consequently are not then favored by selection, and nevertheless do persist and become useful adaptations. Darwin foresaw this objection to his theory and attempted to answer it, but without complete success. A related objection is the claim that many striking and complex characters of animals have no selective value at any time and so cannot be favored by selection. This, again, would merely mean that selection is not the only influence in evolution, and it has no special bearing on adaptation, as such characters are supposed to be neither adaptive nor inadaptive but merely nonadaptive.

Neither of these latter objections proved to be very serious in the light of increased knowledge of the nature and extent of adaptive advantage, but there was another objection that proved to be very serious indeed. After studying numerous lineages among fossils, some paleontologists concluded that evolution tends to proceed in straight lines, that the evolutionary direction of these lines is unswerving even when it has no (evident) adaptive advantage, that it may even run counter to adaptation, and that a trend initially adaptive may continue to such a point that it becomes inadaptive and perhaps fatally disadvantageous.

It was objections of this sort that led many students to believe that attempts at mechanistic interpretations of evolution, and particularly of adaptation, had failed and were foredoomed to such failure. If characters arise regardless of adaptive significance but nevertheless do, for the most part, come to fill a need, if they do not arise at random but in regular progression, or if they progress steadily in straight lines as if toward a goal—then, it was felt, there must be some guiding force controlling evolution, quite possibly a purposeful force, but this force cannot be found in the environment or in the physical laws of nature. It must either be a force peculiar to life, and to this extent nonphysical, or it must reside on some supernatural, or at least extranatural, plane. Such a force was repeatedly postulated by students disillusioned with both Darwinism and Lamarckism. Driesch called it "entelechy"; Bergson called it the *élan vital;* Osborn called it "aristogenesis" (among other things); Teilhard called it "noögenesis." Others call it other names, and some are content to postulate such a force without naming it.

It would certainly be a mistake merely to dismiss these views with a smile or to ridicule them. Their proponents were (and are) profound and able students. Yet in essence what they are all saying is little more than "The cause of evolution is the force that causes evolution." Attempts to define this force usually amount to no more than a description of the way these students suppose evolution to occur, which really leads to no comprehension of how or why it occurs in this way—if, indeed, it does, which is also a decidedly debatable point. As Julian Huxley has remarked, ascribing evolution to an *élan vital* is like "explaining" the movement of a train by an *élan locomotif.* It may be added that "definitions" of these various names for the supposed nonmechanistic causes of evolution generally sound like defining the *élan locomotif* as "that force which causes a train to travel undeviatingly from one station to its goal in another station."

A word must be said, also, on those theories of this general class that go to the extreme of assigning a supernatural cause to evolution, not in the sense of divine establishment of the natural laws of evolution, a belief held by many evolutionists who nevertheless seek to discover these mechanistic laws, but in the sense that the laws themselves are supernatural. Among these views, that of Broom is extreme and yet representative of a trend of thought usually less elaborated and less clearly or frankly expressed. Broom maintained that there is "some spiritual power which has planned and directed evolution" and that below this there are other spiritual agencies, some good and some evil, which in turn direct "partly intelligent" inferior spiritual agencies directly associated with the various animals and plants.

If any of these views are true, we cannot hope to explain "the eternal problem of adaptation," as Osborn called it. They postulate that the purposive element in evolution involves some force that is quite outside the sphere of scientific investigation. We can see what it does but we cannot and need not determine how or why it does this. In fact, these ideas are metaphysical and not scientific, even though this is often denied by their supporters. One cannot but suspect conscious humor when Teilhard (a trained theologian, as it happens, in addition to being an able paleontologist) states in the midst of one of the most subtly metaphysical discussions of evolution known to me, "In this, note well, nothing metaphysical."

Of course an explanation might be metaphysical and nevertheless true. It is, however, an obvious lesson from the history of scientific progress that in science one should never accept a metaphysical explanation if a physical explanation is possible or, indeed, conceivable. In some cases these theories were clearly born of despair and faintness in the search, an emotional state with which we must sympathize but which we should surely seek to avoid in ourselves. Wallace, who independently from Darwin arrived at the principle of natural selection, finally became disillusioned, along with many of his contemporaries, as to the ability of this principle to provide an adequate explanation of evolution. He concluded that "a superior intelligence" must also be involved. Commenting on this after Wallace's death, Osborn wrote of "transformations which become more and more mysterious the more we study them, although we may not join with this master in his appeal to an organizing and directing supernatural principle." Yet Osborn himself, in the course of the twenty-two years that then remained of his long and fruitful life, came to expound evolution as the result of an organizing and directing principle which he refused to call supernatural but for which he found no naturalistic basis or explanation.

The rise of modern knowledge of heredity in the early years of this century brought with it an attempt to dispose of this troublesome, eternal problem by abolishing it, an astonishing solution as seen in retrospect and yet one that held sway among the most advanced biological theorists for almost a generation. It was found that new characters and new types of plants and animals can arise quite suddenly by some change within the hereditary substance in the germ cells. As far as is known, these changes, mutations of various sorts, arise quite at random. The geneticists, fervently exploring on the frontier of this newest offspring among the biological sciences, proposed on this basis to do away with all the musty problems of purpose or adaptation by simply denying their existence. New types of plants and animals arise at random, and that is all there is to it. No adaptation and, of course, no purpose.

The geneticists quickly took a great step forward in the attack on one very essential phase of the problem of adaptation, the question of the nature and origin of hereditary variations. Lamarck faced this question more squarely than almost any later student of the nineteenth century, and he proposed an answer, but his answer was wrong, as the geneticists, among others, have clearly shown. On the whole, Darwin avoided the question. It is doubtful whether even a partial answer was practicable in his day, and to this extent Darwin's study of adaptation was premature, but his evasion or glossing over of the question was a defect in his theory. It led to some confusion among his followers and to a reaction against this concept during the next half-century. The vitalists met the question in a still less satisfactory way, by relegating it to the realm of the unknowable or by self-deceiving naming fallacies.

The geneticists have observed and isolated nascent variations (mutations of various sorts), and they have found out the more essential facts about how these are passed on, combined or separated, and expressed in later generations. Beyond these processes there are other, deeper mysteries into which the geneticists have barely penetrated, if at all—all steps toward greater knowledge bring us newly to the threshold of more profound mystery. But modern geneticists have supplied what seems to have been the last lacking basic information necessary for an explanation of evolution essentially complete on, at least, the descriptive level. Their own attempts to apply this information to evolutionary theory were not, at first, fruitful, although they later became so. In earlier years of the present century most geneticists worked in (mainly self-imposed) isolation from the paleontologists, systematists, comparative anatomists, and others who had accumulated a vast body of data on the course and on the results of evolution. As is so likely to happen when great new discoveries are made, some of these geneticists were overenthusiastically sure that their discovery quite superseded all that had gone before and was the sole and whole key to evolution. In no other way can one understand their proposed solution of the problem of adaptation by saying that there is no such thing as adaptation.

Adaptation does exist and so does purpose in nature, if we define "purpose" as the opposite of randomness, as a causal and not a merely accidental relationship between structure and function, without necessarily invoking a conscious purposeful agency. Denial of this does violence to the most elementary principles of rational thought. Look again, with Sir Charles Bell, at your own hand, manipulate the fingers, think of the intricacy of the combination of bones, muscles, tendons, blood vessels, and nerves in this mechanism and of the complexity and delicacy of its conscious control. How can one estimate the improbability that such a structure arose by sheer accident, or by any continued series of accidents short of infinity? The example is simple but it is convincing in its very simplicity. Consider, then, all the other structures and organs of man and of all the other animals and plants and the way in which these function and serve the many special needs of each type of organism. Consider, too, the interrelationship of these organisms, the dependence of animals on plant photosynthesis, the social groups of ants, the fertilization of flowers by insects, and the myriad other adaptive ecological relationships. Whatever improbability you assigned to the random origin of your hand, this might be multiplied by a billion billion to express the improbability that nature as a whole is the result of any sequence of accidental and random events. It is necessary to agree with Julian Huxley that "to produce such adapted types by chance recombination . . . would require a total assemblage of organisms that would more than fill the universe, and overrun astronomical time."

Although some geneticists really were, for a time, so naive as simply to deny the reality of the problem of adaptation, it would be unfair to imply that this opinion became universal or endured long among them. Acceptance of random mutation as the mechanism of evolutionary change and of the fact that adaptation—or something that looks singularly like it—is well-nigh universal in nature led to emphasis of another evolutionary principle, that of preadaptation. The essence of this principle is that structural or physiological peculiarities, although arising at random, may prove to be suitable or useful in the environment of the organism or in some other environment available to it. For instance, animals living in fresh water may by mutation become tolerant of salty water and then spread into brackish streams or into the sea because they now can; or animals living on a forest floor may undergo mutations enabling them to climb and may then take to the trees for refuge. This is a sort of reversal of the old idea of apparently or really purposeful adaptation in, for instance, the Lamarckian sense. The "adaptation" (it should probably be in quotes when used in this way) comes first and its use follows. According to an extreme version of this theory, all strongly distinctive types of animals originated as sports, or "hopeful monsters," as Goldschmidt calls them, that happened to find a practicable way of life adapted to their peculiarities, rather than originating by any process that adapted them to peculiarities of the environment.

This is an interesting and important theory. There is little doubt that preadaptation does really occur, although rarely if ever in the extreme form supported by Goldschmidt. We shall see that preadaptation, with some expansion and modification of its significance, must be accepted as one of the general factors in the synthetic theory of evolution. Earlier opinions that random preadaptation is an adequate explanation of adaptation were, however, quite unjustified. Objections already expressed to accepting any explanation of adaptation as a result solely of random, accidental, or unsystematic processes apply with equal force to preadaptation in this relatively crude form and without any additional postulate involving a systematic guiding influence and providing for pseudo- (if not really) purposeful progressive increase in adaptation.

At this stage in the inquiry, the situation seems almost hopelessly confused. This was the position only a few years ago, with many special theories ardently supported, although to a dispassionate observer none would have seemed adequate and some would have seemed absurd. Some students, a very small minority, denied the existence of adaptation. A host of other students had, however, demonstrated really beyond any doubt that adaptation does exist and that a great majority of the characters of animals and plants, although not all their characters, must be considered as definitely related to, and requisite for, their particular ways of life. For some this was merely an inexplicable fact; these students were few, because scientists rarely are psychologically capable of accepting a phenomenon as a fact and also accepting it as inexplicable. It is well for the progress of science that this is so, but it has led to many premature and to some bizarre theories.

Die-hard Neo-Lamarckians still maintained that adaptation is explicable as the result of the effects of the environment on the organism and of the organism's own efforts to cope with the environment. Die-hard Neo-Darwinians still maintained that the removal of misfits by natural selection is the whole story. The latest comers, the geneticists, were in many if not in most cases convinced that the only adaptation is preadaptation. Besides these three main mechanistic schools, there were many who supported a wide variety of covertly or overtly nonmechanistic views. In part they were reacting in a natural way against the excesses and deficiencies of each of the mechanistic schools. Rationally unable to accept one or the other of these, they were unwilling to leave adaptation unexplained and sought refuge in nonscientific or metascientific theories. In other cases they were influenced by, or returning to, the preevolutionary and long antievolutionary concept of adaptation as literally purposeful, but, as a rule, they dodged the question of who or what formulated the purpose.

The story of the elephant and the blind men, who argued bitterly over the nature of the beast only part of which was known to each of them, is old and trite. It has become trite because it is so often apt, and it is very apt here. Each of the diverse schools of thought about adaptation knew or emphasized only part of the pattern. Each tried to complete this pattern as a reconstruction from a part only and each bitterly rejected the reconstructions of the other schools, based on other parts. The Neo-Lamarckians knew and overemphasized the fact that adaptation is pervasive in nature and essentially purposeful in aspect, as if the environment had forced and the organism had sought adaptation. The Neo-Darwinians knew and overemphasized the fact that the more or less adaptive status of variations is influential in determining the parentage of a following generation. The geneticists knew and overemphasized the fact that new hereditary variants arise abruptly and, as far as we know and as far as adaptive status is concerned, at random. The various non- and antimechanists knew and overemphasized the fact that adaptation is usually an essentially directional, progressive, sustained, and nonrandom process.

Once the search has been summarized in this way and its outcome expressed in these words, it becomes obvious how to seek further for an answer to the problem of plan and purpose in nature, or of adaptation, to which the problem is reduced by elimination of some of its metaphysical overtones. What was necessary was synthesis, bringing together the facts and theories of all the schools, accepting those mutually consistent and reciprocally reinforcing, rejecting the inconsistent, inadequately supported, and unnecessary, and building anew on this basis. Such a development, obvious as it now seems, would have been premature until rather recently because essential elements, especially some of those lately supplied by the geneticists, were still lacking, and the synthesis would still have been so incomplete that it would have satisfied none of the participants. When the time was ripe for it, it was still impeded by the extreme specialization that had grown up in biology as in other sciences. Students in the various narrow specialties tended to be ignorant of any field but their own and were sometimes downright hostile toward any other.

The breaking down of these barriers, a necessary preliminary not only for further progress in the study of evolutionary adaptation but also for true comprehension and integration of all fields of biological research, did finally occur. A precise beginning can hardly be designated, but this movement was presaged a generation ago by individual efforts, mostly English and American, to renew, broaden, and modernize the attack on evolutionary problems. More recently it has been put on a wider basis and has become a conscious collaboration between many specialists in different branches of biology.

The result is essentially a new theory of evolution, or a new body of theory, although its basis is a synthesis of previous theories. It has been called Neo-Darwinian because it includes the Darwinian factor of natural selection and excludes the Lamarckian factor of inheritance of acquired characters, but this is a misnomer and is likely to mislead seriously. It involves, and to a considerable extent it grew out of, rehabilitation and restatement of the principle of natural selection in genetical and statistical terms, but its understanding of natural selection is quite different from that of Darwin and still more different from that of the Neo-Darwinians. It also embraces a great deal more than natural selection. It is more complex than any one of the previous theories; their factitious simplicity was a weakness and a cause of their failure. It can hardly be labeled with the name of any one man, and an outstanding characteristic is precisely that it is the product of many minds and has drawn data and principles from many fields. It is most appropriately called simply the synthetic theory. Main features of the synthetic theory have here been discussed in Chapter 4 and elsewhere. In the present chapter some further special consideration is given to elements of the theory bearing most particularly on adaptation and the purposive aspects of evolution.

To elucidate, as far as can be done in a brief summary, its solution of the problem of adaptation, it is well first to restate the problem as it appeared to the opposing teams which, by combining their efforts, produced the synthetic theory. The geneticists had, as they thought and as paleontologists, systematists, and other nongenetical naturalists came to accept, identified the materials for evolution: discrete changes in the hereditary mechanism. This is, in the main, an elaborate organization of chemical substances into units called genes, which are assembled into united groups called chromosomes, which in turn occur in sets one or more of which are present in every cell that is to grow into an organism. These chromosome sets as a whole, with all their constituent parts, determine what the organism will become as it grows. The geneticists found, moreover, that the changes in this mechanism, those changes that are the materials for any evolutionary change, are not predominantly adaptive. On the contrary, they have no particular orientation toward adaptation but are random in this respect. Still worse, this means that the great majority of them are definitely opposed to adaptation, because in any organism the number of possible changes toward increased adaptation or toward new adaptations is very much smaller than the number of possible changes away from adaptation. Moreover, what an organism is like is determined by its whole set of chromosomes. Adaptation rarely requires only a single new mutation; more commonly it requires an entire set coordinated in a new way. That genetic mutations should produce a new sort of organism adapted to an available environment is improbable in so high a degree as to seem almost impossible. Even granting that natural selection would, in Darwinian fashion, weed out the grossly unfit genetic combinations, the fit simply would not arise, or would arise so extremely infrequently as to be altogether exceptional in nature.

On the other hand, the set of facts known to the nongeneticists demonstrates beyond any possible doubt that these almost impossibly improbable genetic combinations are in fact so common in nature as to be nearly universal. The only possible way to reconcile the facts of genetics with the facts of adaptation is to find some force or process in nature that is capable of generating a high degree of improbability, as R. A. Fisher put it, or, in other words, of assuring that an outcome that is genetically extremely improbable will nevertheless become usual.

This force has been identified beyond reasonable doubt, and it turns out to be an old friend, natural selection, but natural selection on a new basis and in a new role, a process subtly but fundamentally different from the natural selection of Darwin or of the Neo-Darwinians. It is not merely the negative process of elimination of the unfit by assuring that they will have fewer offspring than the fit; it is the positive and creative process that was left out of the picture by the Darwinians and that was sought in vain by the Lamarckians, the vitalists, and others.

How natural selection works as a creative process can perhaps best be explained by a very much oversimplified analogy. Suppose that from a pool of all the letters of the alphabet in large, equal abundance you tried to draw simultaneously the letters *c*, *a*, and *t*, in order to achieve a purposeful combination of these into the word "cat." Drawing out three letters at a time and then discarding them if they did not form this useful combination, you obviously would have very little chance of achieving your purpose. You might spend days, weeks, or even years at your task before you finally succeeded. The possible number of combinations of three letters is very large and only one of these is suitable for your purpose. Indeed, you might well never succeed, because you might have drawn all the *c*'s *a*'s, or *t*'s in wrong combinations and discarded them before you succeeded in drawing all three together. But now suppose that every time you draw a *c*, an *a*, or a *t* in a wrong combination, you are allowed to put these desirable letters back in the pool and discard the undesirable letters. Now you are sure of obtaining your result, and your chances of obtaining it quickly are much improved. In time there will be only *c*'s, *a*'s, and *t*'s in the pool, but you probably will have succeeded long before that. Now suppose that in addition to returning *c*'s, *a*'s, and *t*'s to the pool and discarding all other letters, you are allowed to clip together any two of the desirable letters when you happen to draw them at the same time. You will shortly have in the pool a large number of clipped *ca*, *ct*, and *at* combinations plus an also large number of the *t*'s, *a*'s, and *c*'s needed to complete one of these if it is drawn again. Your chances of quickly obtaining the desired result are improved still more, and by these processes you have "generated a high degree of improbability"—you have made it probable that you will quickly achieve the combination *cat*, which was so improbable at the outset. Moreover, you have created something. You did not create the letters *c*, *a*, and *t*, but you have created the word "cat," which did not exist when you started.

Creative natural selection works in a similar but vastly more complicated way. The number of possible combinations and arrangements of genes and chromosomes in a group of organisms, even of relatively simple organisms, is enormous, so enormous that most of them will never occur because the number of organisms in the group is much less than the number of possible genetic systems. Clearly the critics of natural selection, in the old sense, were quite right in concluding that merely eliminating inadaptive combinations would rarely, if ever, insure the appearance of adaptive combinations. A mutation favorable in itself or in some particular combination would quickly be swamped among the inadaptive mutations and would usually be eliminated because it occurred in unfavorable associations. But it has been demonstrated both theoretically and experimentally that selection acts in a positive way tending to increase the percentage of favorable genes in a population, a process analogous to your increasing the proportion of *c*'s, *a*'s, and *t*'s in the alphabet pool. It thus greatly increases the chances not only of favorable single genes but also of favorable hereditary combinations. Moreover, selection also acts on combinations and arrangements of genes. Just as you clipped together the "adaptive" sets of letters *ca, ct,* and *at,* selection prevents adaptive combinations of genes from being dispersed again and increases their frequency in the population, thus promoting the probable development of still more complex, adaptively still more favorable, combinations. It has also been demonstrated, to meet another old criticism of natural selection, that under certain conditions (conditions that do obtain in natural populations that are well adapted) an extremely small selective advantage suffices to insure that favorable genes and genetic systems will be preserved and will increase in frequency.

Another major criticism of natural selection, in the old, Darwinian, sense, has also been well answered by the studies of, particularly, paleontologists and geneticists. This criticism, you will recall, was that evolution exhibits trends that proceed without deviation ("orthogenetically," in paleontological cant) and that may be inadaptive or may proceed well beyond the point of greatest adaptive advantage. In some cases these trends simply did not really exist or were not really the single, straight-line affairs that they had been thought. Many of these straight lines are more a product of scientists' minds than of nature. In other cases, the interpretation of some features of these trends as inadaptive seems to have been due to an inadequate, sometimes a naive, conception of adaptation and of heredity. One factor here is that the hereditary mechanism makes it quite possible for an unfavorable trend to become associated with a favorable trend and to be carried on with it as long as the balance continues favorable. There are some remarkable complexities involved in this subject, but it seems reasonably clear that long-continued trends in general are kept going by natural selection.

The action of selection, understood in this newer way, results in the appearance and spread of genetic systems and therefore of sorts of organisms that would never have existed under the uncontrolled influence of mutation and random recombination of the elements of heredity. In this sense, although it does not create the raw materials, the mutations, natural selection is definitely creative. It creates the most important product of all, the integrated organism. Builders do not make bricks, but they create houses, and the bricks are not adapted into a use until they are assembled into houses. Natural selection is a builder that uses mutations as bricks, and its constructions are adaptive. The figure of speech would be still more apt if builders had to find their bricks in piles of rubble in which most of the pieces were inappropriate for their purpose.

The whole process and its results are much more intricate than is apparent from this simplified explanation, and natural selection is by no means the only process involved in evolution or even in adaptation. Creative natural selection is the directive, pseudopurposive factor back of adaptation, but it is not always the decisive factor in evolution and it never acts alone. Other factors explain, for instance, why not all characters of organisms are adaptive, why not all newly evolved sorts of organisms are adaptively superior to their ancestors, and why organisms adapted to essentially the same activities and environmental conditions may nevertheless be strikingly different in many respects.

Among the other processes of evolution, one particularly pertinent to the present theme requires further brief comment, and that is preadaptation. Preadaptation commonly occurs in evolution and is sometimes of crucial importance, but it is not as universal as was believed by its earlier proponents, and its role and the factors underlying it are rather differently understood in the synthetic explanation of evolution. It now appears that preadaptation may arise in several quite distinct ways, none of which is wholly random or spontaneous, as all preadaptation was once thought to be. Occasionally one mutation, using the word in a broad sense, may produce a relatively great difference from the parents in the organism affected by it, and this difference may be adaptively favorable for available new habits or environments. This is, in a sense, preadaptation, but in the same sense any useful mutation is preadaptive, and this application of the term is not enlightening. Once such a mutation has arisen, it is simply another bit of material for natural selection and its status is the same as any other mutation except for its larger size, in terms of structural change. Large or small single-step adaptations, as Julian Huxley calls them, seem, however, to be of relatively little importance in evolution as a whole. They explain occasional evolutionary events, but the usual course of adaptation is slower and more cumulative.

One type of preadaptation, speaking more strictly, may occur when selection becomes relatively ineffective and the frequencies of the various genes in a population tend to drift, aimlessly as it were, without definite control by selection. This was referred to as "sampling error" in Chapter 4. In most cases the changes brought about in this way are inadaptive, and they are usually a prelude to extinction, but occasionally they are preadaptive. Then they may facilitate a relatively abrupt and radical change in habits and environment, a change of the sort that I have called "quantum evolution." This is probably the nearest thing to purely random preadaptation in the old sense, but even this cannot be considered wholly random. The structure from which such developments depart owed its integration to (nonrandom) selection, and the new grade of structure must, if it survives, in turn be integrated by selection which will act with particular force on such groups.

Another, and more common sort of preadaptation arises when,

for instance, a selectively controlled adaptive structure develops to the point where its use in a new way becomes possible. Then it is preadaptive with respect to its new use, and selection will then direct evolution in this new direction if the new use is advantageous. This sort of preadaptation is not random at any point. It is always directional and directed, but the direction changes. Such occurrences demonstrate how the direction of evolution can change under the influence of selection even though the environment remains essentially constant.

Adaptation by natural selection as a creative process and preadaptation in the special senses just explained are the answers of the synthetic theory of evolution to the problem of plan and purpose in nature. Of course much work remains to be done, many details to be filled in, and many parts of the process to be more clearly understood, but it seems to me and to many others that here, at last, is the basis for a complete and sound solution of this old and troublesome problem.

Adaptation is real, and it is achieved by a progressive and directed process. The process is wholly natural in its operation. This natural process achieves the aspect of purpose without the intervention of a purposer, and it has produced a vast plan without the concurrent action of a planner. It may be that the initiation of the process and the physical laws under which it functions had a Purposer and that this mechanistic way of achieving a plan is the instrument of a Planner—of this still deeper problem the scientist, as scientist, cannot speak.

CHAPTER ELEVEN

Evolutionary Theology:
The New Mysticism

THE premise of the *Bridgewater Treatises* was that God's works reveal His existence and nature. His attributes were stated *before-hand* to include power, wisdom, and goodness. Expectations were entirely fulfilled: those were, indeed, the qualities that the Bridge-water authors found manifest in creation. At the dawn of the Victorian era, those authors sought and found an eminently suitable Victorian God, the same from inspection of the world of phenomena as from divine revelation and from inspired tradition. That harmony had not always existed, nor was it long to endure.

There could not primitively be any conflict between percep-tion of the natural and conception of the supernatural. No real distinction was made between the natural and the supernatural, which were simply united in a single and necessarily harmonious world picture. A creed involving a motionless sun and a moving spherical earth would have been rejected out of hand—indeed would never have been proposed—when any fool could see the sun moving over the fixed flat face of the earth. But the example shows the dangers inherent for both science and religion when they become somewhat more sophisticated. The earth is not in fact flat or mo-tionless, and objective inquiry was bound to correct those errors sooner or later. Then it became necessary either to reject the primitive unified world picture or to give up the rational investigation of the material world.

The conflict between science and religion has a single and simple cause. It is the designation as religiously canonical of any conception of the material world open to scientific investigation. That is a basis for conflict even when religion and science happen to agree as to the material facts. The religious canon (if normally designated as such) demands absolute acceptance not subject to test or revision. Science necessarily rejects certainty and predicates acceptance on objective testing and the possibility of continual revision. As a matter of fact, most of the dogmatic religions have exhibited a perverse talent for taking the wrong side on the most important concepts of the material universe, from the structure of the solar system to the origin of man. The result has been constant turmoil for many centuries, and the turmoil will continue as long as religious canons prejudge scientific questions.

If we are willing to define God as the Author of the cosmos, with no further a priori qualifications, then there must indeed be something in the Bridgewater premise: the study of nature should truly bear on the attributes of God. But the scientific revelation is gradual and changing. It makes no pretension of absolute knowledge and shows no sign of reaching finality. If the "rational" and "natural" theology of Paley, the Bridgewater authors, and other worthies of that and earlier periods had really been rational and natural, it would have had to recognize that revelation is continuously modified and never complete. For most of the established religions, such recognition would destroy their claim to authority and would therefore be unthinkable. Moreover, in the course of the nineteenth century it began to appear that the God revealed by the study of nature was not after all the comfortable Victorian God. Natural theology was making the wrong kind of revelation.

Reactions to that situation have been extremely varied. Some would have liked to reject all the findings of science, but that has become impossible for anyone even moderately sane and informed. Many sects have reluctantly undergone a form of intellectual pruning, slowly abandoning beliefs now in clear contradiction with scientific observation. This can be done without also abandoning the pose of authority by declaring that the former canonical belief now abandoned was never *really* infallible dogma—and that beliefs not yet abandoned remain true as always. Some individuals have given up canons and dogmas altogether in favor of the continuing revelation of nature. A commoner reaction has, however, been quite the opposite, to jettison natural theology and the idea that the study of nature reveals attributes of God. Knowledge of God is then sought not by mediation through the outer and material world but more directly through the inner world of emotion, intuition, and mystic communion.

I am no theologian, natural or otherwise, and I am unwilling to pose as one. Here I am also not directly or primarily concerned with the conflict between science and religion, but with a peculiar, related interaction of the two. The path of religious intuition or of mystic communion need not have conscious connection with the material world, and indeed this is one of several ways in which the conflict with science can be successfully resolved. However, for some scientists there has seemed to be such a connection, and then some reconciliation or fusion of the two conceptual schemes has been sought. The relationships envisioned vary greatly. For Julian Huxley (*Religion Without Revelation*) theology itself becomes a subject for scientific investigation and religious emotion is a psychological fact centering on direct experience of sacredness in the universe. For Teilhard de Chardin the mystic conviction is overwhelming and primary; it is the premise for all interpretation and overrides any consideration of objective science. Huxley and Teilhard could hardly differ more as regards theories of evolution, attitudes toward science, and conclusions as to theology, but they both have proposed systems in which, in quite different ways and proportions, science and mysticism are involved.

It is that phenomenon that I shall discuss in the present chapter. This will be done mainly by reference to the works of three biologists whom I would call new mystics: Pierre Lecomte du Noüy, Edmund W. Sinnott, and P. Teilhard de Chardin. They are nearly alone, or at the least have been most eloquent, in expression of varieties of evolutionary theology or theological evolution. All three are both finalists and vitalists of sorts and so those minority schools, given short shrift elsewhere in this book, will have some notice. All agree with other finalists and vitalists that a naturalistic theory of evolution has not been and cannot be achieved. All make appeals to the supernatural, in these three cases to different forms of dogmatic Christianity, and all advocate concepts or theories of evolution in which the approaches of science and religion are confused. With varying degrees of overtness, all make some claims that their religious conclusions are derived from scientific premises and interpretations. That aspect has led in some instances to popularity among those piously concerned but not cognizant of the actual issues or capable of judging the theories. One suspects that these best sellers, like many in other fields, were more bought than read.

Specifically, Lecomte du Noüy's book *Human Destiny* was a best seller for months in 1947, with accolades from *Reader's Digest,* from numerous clergymen and journalists, and, indeed, from some scientists. It is difficult now to comprehend the excitement over that book, which was expected to "change the whole direction of scientific thought since the Renaissance." In fact it turned out to have no perceptible influence whatever on scientific thought. Even leading studies of the same character, that is, with a religious approach to consideration of evolution, have almost consistently ignored it. That would be sufficient excuse to omit consideration of it here, too, but there is some interest in the very fact that the book has been virtually forgotten after creating such a furor only a few years ago. And it does illustrate one variant of the new mysticism.

Lecomte du Noüy differentiates between *evolution,* which is limited to the long lineage leading to man, and *transformation,* which is organic change in all other lines or in general. The supposed distinction, perhaps due to Bergson but with even earlier antecedents in other terms, has been adopted by several French philosophers. It has caused much misunderstanding among non-French students, for whom *evolution* is the term for any and all organic progression. In fact the distinction is purely arbitrary or metaphysical, without any objective and scientific basis. There is no more reason for a special name for change eventuating in man than for change eventuating in any other species.

Lecomte du Noüy then goes on to say that transformation is caused by mutation, adaptation, and natural selection. He makes no clear distinction between somatic adaptation of an individual and genetic adaptation of a species, but assumes that one leads to the other. He takes the inheritance of acquired characters for granted, in part because he confuses induced mutations and chromosomal aberrations with acquired characters. In fact and in the sense required by the theory, the two have no relationship and acquired characters certainly are not heritable. Evolution, that is, the origin of man, is said to be affected by the same factors but not explained by them. It is not adaptive, for adaptation is conceived of as taking organisms *out of* the tenuous main line of evolution. Perhaps there is here a dim echo of Lamarck. As was noted in Chapter 3, Lamarck also considered adaptation as divergence from the main line of evolution, a line leading inevitably and only to man. (Lecomte du Noüy seems to have been unaware of this resemblance in the theories; he refers to Lamarck only in connection with adaptation and acquired characters.)

As to what does cause progression in the main line of evolution, Lecomte du Noüy is no clearer and no more successful than Lamarck. In fact, as far as he is concerned, nothing was learned about evolution between 1809 (publication of *Philosophie Zoologique*) and 1947 (*Human Destiny*). Evolution, as opposed to transformation, occurred because it was willed. It is, indeed, taken right out of the realm of science and ascribed to the inscrutable Will of God. This view—it cannot be called a theory because it is nonscientific, untested, and untestable—is called "telefinalism," a garbling of Latin and Greek intended to unite and modify concepts of teleology and of finalism. Lecomte du Noüy repeatedly calls telefinalism an "explanation," but it is simply a word applied to professed ignorance. In what sense can it possibly explain?

There is no positive connection at all between this word and scientific procedures, but Lecomte du Noüy attempts to supply a negative connection or argument. Boiled down to its bare bones, the argument is simply that in his opinion evolution cannot be explained scientifically and we might as well cover inability to explain it with the word "telefinalism." In support of his opinion Lecomte du Noüy argues that evolution cannot possibly be due entirely to chance and that it must therefore be due to divine Will. This argument, always advanced by finalists in one form or another, is wildly illogical. No one, at least certainly no scientist, ever supposed that *any* natural event occurs entirely by chance. Ours is, we believe, a lawful universe. The issue is whether the explanation of evolution is by natural laws and processes or whether it continually involves the supernatural, hence evades orderly causation and becomes merely inexplicable. Neither view does or can ascribe evolution simply to chance, and the finalists' argument is merely a *non sequitur*. Lecomte du Noüy does not even notice that two of the materialistic causes of transformation that he accepts, Lamarckian adaptation (which does not really occur, but he says it does) and Darwinian natural selection (which does occur), are precisely such antichance factors as he insists cannot be found except in the supernatural.

Lecomte du Noüy's explicit arguments from probability are more involved but are equally fallacious. His treatise begins with and rests throughout on a mathematical "proof" that life could not arise by natural means. The fallacy of the "proof" no longer needs detailed discussion, because, only a few years after he wrote, an essential step that he had "proved impossible" was experimentally performed. It is now accessible to any schoolboy. Another part of his supposed evidence is an outline of the history of life, from which he concludes that all proceeds as if by Will and is quite inexplicable in natural terms. This account is so full of errors and contradictions, so completely unreliable, that it proves nothing beyond the fact that its author was not a student of this subject.

What I have said is sufficient to explain why *Human Destiny* has had no influence on the scientific study of evolution. It does not explain why a scientist who was not only respectable but also eminent wrote such an unscientific work or why it had its day of popular acclaim. I am myself baffled by that phenomenon, but there are some clear hints. A considerable part of *Human Destiny* consists of pious moral and ethical discourse, lofty in tone and admirable in intention. The author (whom I did not know) clearly was a devout Christian and, more than that, a man of good will. His original scientific research was in physiology and biophysics and had no direct connection with evolutionary biology. He saw that evolution must of course be accepted, but he also saw that "materialistic" (I would say naturalistic and scientific) views on evolution were disturbing to the faith of some fellow Christians, and, I suspect, also to his own faith. With the highest moral and ethical motives, he approached the subject with that *parti pris*—not to learn what it could tell him but to reassure himself that it could tell him nothing shocking or new about God. Under the guise of science this became a work of mystical exegesis. "Seek and ye shall find" has more than one application. The popularity of the book was among those who thought or merely hoped that Lecomte du Noüy had found something they also sought.

Edmund W. Sinnott has reached conclusions similar to those of Lecomte du Noüy in certain fundamentals, but he has done so in quite a distinct way. His interests are different until they reach the religious, theological, mystical level, and his approach develops more directly from his real biological competence. Sinnott, now retired, is one of the most distinguished American biologists, with a long career in research, teaching, and administration at Harvard, Connecticut, Columbia, and Yale. His research has centered on the genetics and morphogenesis of plants. The views pertinent to our present enquiry were developed especially in three rather short books: *Cell and Psyche* (Chapel Hill: University of North Carolina Press, 1950); *Two Roads to Truth* (New York: Viking, 1953); and *The Biology of the Spirit* (New York: Viking, 1955).

Sinnott is concerned above all with the relationship between the material body of an organism and instinct, mind, psyche, or spirit (different levels or manifestations of the same thing). He starts with some of the facts of development in organisms, the truly extraordinary way in which individual organization is achieved.

From spore, egg, seed the development to ultimate intricacy proceeds almost without deviation and even, in some instances and to some extent, with correction of deviations that may be imposed. Biological self-regulation is a fact, which no biologist would dream of questioning whatever his philosophy or religion might be. Sinnott points out that this self-regulation has not been fully explained, and while this is also correct to the point of being obvious, here an inchoate element of mysticism already may be suspected. It is the most general and yet the most completely invalid argument for vitalism, finalism, and ultimately mysticism that what *has not been* explained in physical (that is, natural) terms *cannot be*. Sinnott does not quite fall into that fallacy, because he does admit the possibility that a physical explanation will be found. Yet for all his proper caution, he does not really believe so: his pursuit of the subject brings him finally face to face not with DNA but with God.

Sinnott strongly emphasizes that biological regulation is "purposive" and "goal-seeking." He points out that the so-called goal may be merely the end of any sequence of events and that "purposive" may be redefined to mean no more than "directional." Yet, again, the insistent use of these terms prejudices the inquiry, or anticipates a conclusion that something more than or different from physical causation is involved. Sinnott then turns to behavior and behind behavior to the mind, which he considers as also purposeful and goal-directed. He concludes—and this is *his* goal at the scientific level—that the two purposive and goal-seeking phenomena, development and mind, are different aspects of the same thing, that is, of biological regulation. Not in the books cited but in a later comment [*Science,* CXL (1963), p. 764] Sinnott admits that this "obviously does not 'explain' mind," but goes on to claim that "it makes a statement of the problem that may be useful in determining the relationships of biology to psychology."

In fact does it do even that? I think not. Merely pointing out that two phenomena have some aspect that can be described in the same terms does not necessarily point in a useful way to a real relationship between them. I think that it does not in this case. It would if there were demonstrably and testably a causal connection explaining the descriptive similarity. In this case we have no reason to think that purposefulness (to call it that) in development and in mind has the same or a similar basis. On the contrary, we have every reason to think that it does not. The processes of development and of mind are quite obviously different, although it is equally obvious that they are not independent. Mind, like, say, digestion but in a different and much more intricate way, is something that the body *does*. In that sense it depends on bodily structure and development, but it is a different *sort* of phenomenon. To say that it is a different aspect of the same thing is a false analogy, almost a word game comparable to some of Butler's. One must, indeed, question even the descriptive validity of the analogy. It is obvious, if only by introspection, that *some* behavior and mental activity are purposeful in the fullest sense of the word. It does not follow and certainly is not an established conclusion that all mental activity is fundamentally purposeful unless, with banal and confused semantics, you equate "purpose" and "adaptation" (as I think Sinnott sometimes does). That "all ideas at first were purposes" (*The Biology of the Spirit*, p. 49) is at least debatable.

So far I have been following in friendly but critical fashion definitely scientific aspects of Sinnott's thesis. This has been preliminary to consideration of his religious and finally mystical views, which he relates to the scientific thesis although I find the connection quite tenuous at times. Sinnott does not admit those views to be mystical but repeatedly acknowledges that they will seem so to his colleagues. He developed them in two different ways in the last two of his three cited books.

In *Two Roads to Truth*, Sinnott maintains that the world's confusion results from a fundamental opposition between reason [science] and spirit [religion]. He rejects the ideas that either one, alone, suffices or that one can lead to the other (although in his next book he was to connect them more closely). Specifically, knowledge of the spirit cannot come from reason but from insight, intuition, and mysticism (here explicit: pp. 54–55). Sinnott maintains that both reason and spirit give valid knowledge of the universe and that both must be accepted, each in its own terms, even though "concessions on both sides will be required" (p. 25). (I must confess to distaste for the idea that scientists should compromise reason, or that they need to do so in order to accept religion.) The rest of the book is mostly devoted to delimiting the fields traversed by the "two roads" and expounding the concessions that must be made to solve the conflict. All this is fascinating and provocative and much of it strikes me as wise, but it is not further pertinent to just the present topic, being rather a separation than an interaction of science and religion.

It is *The Biology of the Spirit* that goes on to mysticism and reaches a variety of evolutionary theology from the hypothesis that bodily development and mind are different aspects of biological regulation. As I have suggested, even that hypothesis might predispose toward an eventual mysticism—a wary dipping of the toe, so to speak, into the divine spring. A restatement (on p. 104) prepares for the final plunge: "The concept we have here been advocating . . . [is] not . . . committed to either body or mind as the essential reality, since it regards *both* as derived from something deeper than either. What this something is, which thus becomes the source of body, mind, and spirit, is the final problem." Sinnott then rejects natural selection, mechanism, some forms of vitalism (although he is a vitalist himself), and organism almost out of hand. He does not make even a gesture toward the overwhelmingly abundant real evidence for a naturalistic explanation of evolution. All is mystery. There is nothing definite and scientific to be said. But (p. 118) "there is something in life that produces harmony and pattern in a material system and that keeps it moving toward a definite end." (Incidentally, that sentence alone is sufficient evidence that Sinnott's thesis really is both vitalistic and finalistic, although he has sometimes argued the matter of definition.)

The parable of an orchestra then indicates that the "something" that produces pattern in life and moves it toward a goal is an unseen composer. By implication, the unseen composer either is or is responsible for biological regulation—this looking back over our shoulder to the scientific preamble. Looking forward to the final chapter, we find the unseen composer becoming the Principle of Organization (now capitalized), which is first (p. 162) an attribute of God and then (p. 164) becomes God Himself. Thus by a different and I must say a much more interesting and better constructed road we have arrived where we meet Lecomte du Noüy and telefinalism. What have we accomplished? The scientific problem was to explain biological regulation, and we have decided that its cause is the principle of organization. That is not enlightening! (This is the *élan locomotif* fallacy again.) And beyond that all is, I humbly submit, *non sequitur*. I do not deny and in fact I rather envy Sinnott's revelation, but it is a revelation private to him. With respect to the foregoing *biological* considerations, it is prejudice and not conclusion.

There is another and more overt prejudice in Sinnott's books, especially the last one, that I must protest before turning to other matters. Sinnott constantly decries the "materialistic" or, what he seems to think is the same thing, "mechanistic" scientist. That nasty scientist is a communist, someone who "worships matter" and who thinks man is "a machine, an automaton, without freedom, responsibility, or worth." This diabolic enemy is of course a straw man. The materialist does not worship matter. Sinnott's real objection is that he does not worship anything, or perhaps that he thinks that worship is appropriate to a church and not to a laboratory. He does not consider himself or anyone else a worthless automaton. He is, more often than not, philosophically naturalistic rather than strictly materialistic, that is, he disavows special knowledge of the supernatural and eschews its injection (as literally *Deus ex machina!*) into his objective investigation of the universe. He is, in fact, the great majority of scientists who are leading devoted and, frequently, God-fearing lives.

The Bridgewater authors found a Victorian God. Sinnott finds a liberal American Protestant one—quite definitely Protestant, for Sinnott is outspokenly anti-Catholic on the reasonable grounds that Roman Catholicism is averse to some aspects of science and also is obviously not going to accept Sinnott's religious views. That gives an added interest to consideration of our third and most mystical evolutionary theologian, a deeply committed Catholic although not exactly an orthodox one: Pierre Teilhard de Chardin.

Father Teilhard was the only one of our three mystics who was not an amateur as a theologian. He learned philosophy and theology in Jesuit institutions, entered the Society of Jesus at the age of eighteen, and was ordained a priest at the age of thirty-one (in 1912). He was also professional as a scientist, having studied and practiced paleontology and geology at the Muséum d'Histoire Naturelle (Jardin des Plantes, Paris) and also having taken a doctorate in those fields at the Sorbonne (1922). Unlike Lecomte du Noüy and Sinnott, his scientific training and research were precisely on the subject of evolution, and that, combined with long training in theology, gave him altogether unique qualifications. Already in the 1910's he was developing an evolutionary *mystique.* He insisted throughout his life that his was the one true interpretation of Roman Catholicism, but his superiors in the Society and the Church did not agree. By the 1920's his unorthodoxy had become a scandal; he was forbidden to teach and was quietly shipped off to effective exile in China. After World War II he could return to France, but he was still forbidden to teach or to publish on theological subjects. He finally moved to the United States, where he died in 1955.

Teilhard observed his vow of obedience to the extent of limiting open publication of his views, but they were made widely known to his friends. (I was happy to be of that large number.) He left a great mass of manuscripts in such a way as to assure their eventual publication. He was hardly in his grave when his two most extensive discussions of evolutionary mysticism were published in Paris: *Le Groupe Zoologique Humain* (Éditions Albin Michel, 1955) and *Le Phénomène Humain* (Éditions du Seuil, 1955). The latter publisher continues to bring out volumes of Teilhard's works, announced as a complete collection but until now entirely philosophical and theological, not scientific. Volumes of letters, reminiscences, and biography are also appearing. Teilhardian societies have been formed, and there is every aspect of a developing religious cult with Teilhard as its seer or prophet, to use no stronger word. That is in part due to memories of the man himself, for he was extraordinarily likable, with an indescribably warm and sparkling personality combined with an attractive humility on every subject but one. However, the cult is increasingly embraced by people who did not know the man, and it seems that his mysticism has a deep appeal for some hopeful souls under existing circumstances.

Of Teilhard's numerous works, *The Phenomenon of Man* (New York: Harper, 1959) is most pertinent to the present discussion. This is a generally good translation of *Le Phénomène Humain,* written in Peking in 1938–40 and first published in 1955 in Paris as noted above. In what follows, I shall be referring for the most part to the English version of that book. It is, however, part of the background that I have read all of Teilhard's published and some of his unpublished works, both scientific and theological, and that I also conversed with Teilhard at length on these subjects.

In the English version of *The Phenomenon of Man* an introduction by Julian Huxley is substituted for a shorter *avant-propos* to the French edition by N. M. Wildiers, a theologian. The new introduction naturally reflects Huxley's own interests and also the warm sympathy for Teilhard felt by everyone who knew him. It is not so adequate in evaluating the broader intent and nature of Teilhard's thought. That becomes fully evident only in the light of this fact: Teilhard was *primarily* a Christian mystic and only secondarily a scientist.

One of Teilhard's fundamental propositions is that all phenomena must be considered as developing dynamically, that is, in an evolutionary manner, in space-time. "It [i.e., evolution] is a general condition to which all theories, all hypotheses, all systems must bow and which they must satisfy henceforward if they are to be thinkable and true." On this basis, unimpeachable in itself, he reviews briefly the evolution of the cosmos and at greater length that of organisms and of man. He is not concerned with details but with the broadest features of the story as it moves onward. These are traced in a style often delightfully poetic but sometimes syntactically overcomplex and often obscurely metaphorical. The whole process, from dissociated atoms to man, is seen as a gradual progress with two revolutionary turning points among others of less importance: first, and early, the achievement of cellular organization; second, and late, the emergence of true man. Much of the intervening story is envisioned in terms of the succession and frequently the replacement of what Teilhard usually called *"nappes,"* a term difficult to translate that is rendered as "layers" or "grades" in the English version. The *nappe* phenomenon involves the expansion or radiation of groups that had reached new structural and adaptive levels, and it has been extensively discussed by other evolutionists in those or similar terms.

Teilhard's book is not, however, strictly or even mainly concerned with describing the factual course of evolution. That is "the *without*" of things, and the author is concerned rather with "the *within.*" The within is another term for consciousness (the French *"conscience,"* another word without a really precise English equivalent), which in turn implies spontaneity and includes every kind of "psychism." Consciousness, in this sense, is stated to be a completely general characteristic of matter, whether in an individual atom or in man, although in the atom it is less organized and less evident. The origin of the cell was critical because it involved a "psychic mutation" introducing a change in the nature of the state of universal consciousness. The origin of man was again critical because at this stage consciousness became self-consciousness, reflection or thought. Now this as yet highest stage of consciousness begins a concentration or involution that will eventually bring it into complete unity, although without loss of personality in that collective hyper-personal. Then the consciousness of the universe, which will have evolved through man, will become eternally concentrated at the "Omega point," free from the perishable planets and material trammels. The whole process is intended; it is the *purpose* of evolution, planned by the God Who is also the Omega into which consciousness is finally to be concentrated. Mystical Christianity is to be the path or the vehicle to ecstatic union with Omega.

Teilhard's first sentence in *The Phenomenon of Man* is as follows:

If this book is to be properly understood, it must be read not as a work on metaphysics, still less as a sort of theological essay, but purely and simply as a scientific treatise.

In the last chapter (before the epilogue, the postscript, and the appendix) he wrote:

Man will only continue to work and to research so long as he is prompted by a passionate interest. Now this interest is entirely dependent on the conviction, strictly undemonstrable to science, that the universe has a direction and that it could—indeed, if we are faithful, it should—result in some sort of irreversible perfection. Hence comes belief in progress.

But the direction of evolution toward an irreversible perfection is the whole theme, and not merely a philosophical appendage, of the book. Hence we have a book submitted purely as a scientific treatise and yet devoted to a thesis admittedly undemonstrable scientifically. (The word here translated as "belief" is *"foi,"* and the context makes it unmistakable that religious faith is meant.) The anomaly is partly explained by the fact that in this particular manuscript Teilhard did avoid explicit discussion of certain points in dogmatic theology. The origin and fate of the individual soul, Adam and Eve and original sin, and the divinity of Christ, for instance, are all alluded to or allowed for, but only briefly and in veiled terms. In addition the discussion begins as a sort of mystical science and only gradually, almost imperceptibly, becomes mystical religion. Identification of Omega with God is evident from the beginning to anyone already familiar with Teilhard's thought, but in this book it is not made explicit until the epilogue. In others of Teilhard's works, lacking the pretense of being scientific treatises, Omega is discussed in frankly theological-mystical terms.

The sense in which Teilhard's science, and not alone his theology, must be called mystical may be illustrated from an early passage in *The Phenomenon of Man* (Chapter II) that introduces concepts crucially used throughout all that follows. First he makes a distinction between material energy and spiritual energy, and points out that material energy, for instance that derived from bread, is not closely correlated either in intensity or in variety with spiritual energy, for instance that exhibited by human thought. But surely both of these can be described and at least conceivably may be explained in material terms. A reasonable mechanical analogy is provided by a television set, in which the activity and multiplicity of the pictures are not well correlated with the intensity and uniformity of the power in wires and tubes. One could speak of "picture energy" as distinct from "electron energy," but it would be evident that "energy" is not even roughly comparable in the two senses and that the metaphorical terminology obscures rather than promotes understanding of the phenomena. And surely it is further obfuscation when Teilhard goes on to explain:

> We shall assume that, essentially, all energy [i.e., both material and spiritual energy] is physical in nature; but add that in each particular element this fundamental energy is divided into two distinct components: a *tangential energy* which links the element with all others of the same order (that is to say of the same complexity and the same centricity) as itself in the universe; and a *radial energy* which draws it towards ever greater complexity and centricity—in other words forwards.

Alas! That is no better than double talk, from the statement that something *defined* as spiritual is nevertheless *assumed* to be physical onward through the whole discussion.

As to the mechanism of evolution, obviously a, or indeed *the,* crucial point of the scientific part of the inquiry, Teilhard accepted both Darwinism and Neo-Lamarckism as partial factors. He called Darwinism evolution by chance (although natural selection is the only objectively established antichance evolutionary factor) and therefore considered the nonchance Neo-Lamarckian factors more important (although, as he knew, most biologists consider them not merely unimportant but nonexistent). However, he maintained that these and all proposed material mechanisms of evolution are related only to various details of the process. The basic over-all pattern and also the essentially directional elements in its various lineages he ascribed to orthogenesis.

Orthogenesis was variously defined by Teilhard as the "law of controlled complication" which acts "in a predetermined direction," as "definite orientation regularizing the effects of chance in the play of heredity," as "the manifest property of living matter to form a system in which 'terms *succeed each other* experimentally, following the constantly increasing values of centro-complexity,'" or as "*directed* transformation (to whatever degree and under whatever influence 'the direction' may be manifested)." The last definition, which is from a brief manuscript written just before Teilhard's death, is broad enough to include the effects of natural selection, but that was certainly not intended, because Teilhard repeatedly contrasted selection with orthogenesis and indeed usually treated them as complete opposites. Similar imprecision or contradiction in definition is one of the constant problems in the study of the Teilhardian canon.

Indeed these and other usages of the term "orthogenesis" in Teilhard's work seem at first sight to have no explanatory meaning whatever but to be tautological or circular. History is inherently unrepeatable, so that any segment of a historical sequence (such as that of organic evolution) begins with one state and ends with another. It therefore necessarily has a direction of change, and if orthogenesis is merely that direction, it explains nothing and only applies a Greek term to what is obvious without the term. However, when Teilhard says that the direction is "pre-determined" and that there is only one direction—toward greater "centro-complexity," toward Omega, ultimately toward God—then the statement is still not explanatory and is obviously not science, but it is no longer trivial.

Now it is easy enough to show that, although evolution is directional as a historical process must always be, it is multidirectional; when all directions are taken into account, it is erratic and opportunistic. Obviously, since man exists, from primordial cell to man was one of the directions, or rather a variety of them in succession, for there was no such sequence *in a straight line* and therefore literally orthogenetic. Teilhard was well aware of the consensus to that effect, but he brushed it aside and refused to grapple with it in terms of the detailed evidence.

Here we come to the real crux of the whole problem: Which are the premises and which the conclusions? One may start from material evidence and from interpretive probabilities established by tests of hypotheses, that is, from science. Despite the objections of some philosophers and theologians, it is then legitimate to proceed logically from these premises to conclusions regarding the nature of man, of life, or of the universe, even if these conclusions go beyond the realm of science in the strictest sense, and that is not only legitimate but also necessary if science is to have value beyond serving as a base for technology. On the other hand, one may start from premises of pure faith, nonmaterial and nontestable, therefore nonscientific, and proceed to conclusions in the same field of the nature of the material cosmos. It cannot be argued that this approach from metaphysical or religious premises is *ipso facto* illegitimate. It is, however, proper to insist that its conclusions should not be presented as scientific, and that when they are materially testable they should be submitted to that scientific discipline. Gradual recognition of that necessity has been evident in the historical change in the relationships between science and religion.

Teilhard's major premises are in fact religious and, except for the conclusion that evolution has indeed occurred, his conclusions about evolution derive from those premises and not from scientific premises. One cannot object to the piety or mysticism of *The Phenomenon of Man,* but one can object to its initial claim to be a scientific treatise and to the arrangement that puts its real premises briefly, in part obscurely, as a sort of appendage after the conclusions drawn from them. That this really is an inversion of the logic involved is evident from the whole body of Teilhard's philosophical writings and also from the statements, or admissions, made toward the end of this book. A passage indicating that the main thesis of the book is a matter of faith and not scientifically demonstrable has already been quoted. Elsewhere in Teilhard's work there is abundant testimony that his premises were always in Christian faith and especially in his own mystical vision.

That is evident, too, in the complex concept of Omega that is the key to Teilhard's personal religious system. He explained on various occasions that the concept is necessary in order to keep mankind on its job of self-improvement and in order to evade distasteful thoughts of aimlessness and eventual death—worthy but certainly not scientific premises. The following passage from another of his manuscripts may additionally represent this contribution of Teilhard's, also essential in *The Phenomenon of Man* but there even less clear:

> In order to resolve the internal conflict that opposes the innate evanescence of the planets against the necessary irreversibility developed on their surface by planetized life, it is not enough to draw a veil or to recoil. It is a case for radical exorcism of the specter of Death from our horizon.
>
> Very well, is it not that which permits us to form the idea (a corollary, as we have seen, of the mechanism of planetization) that there exists ahead of, or rather at the heart of, the universe, extrapolated along its axis of complexity, a divine center of convergence. Let us call it, to prejudge nothing and to emphasize its synthesizing and personalizing function, the *Omega point*. Let us suppose that from this universal center, this Omega point, there are continuously emitted rays perceptible only, up to now, by those whom we call "mystic souls." Let us further imagine that as mystical sensitivity or permeability increases with planetization, the perception of Omega comes to be more widespread, so that the earth is heated psychically while growing colder physically. Then does it not become conceivable that humanity at the end of its involution and totalization within itself may reach a critical point of maturation at the end of which, leaving behind earth and stars to return slowly to the vanishing mass of primordial energy, it will detach itself psychically from the planet in order to rejoin the Omega point, the only irreversible essence of things?

(Translated by me from an essay written in 1945, published in *L'Avenir de L'Homme,* Paris: Éditions du Seuil, 1959, pp. 155–156.)

That is mysticism at its purest (if not exactly simplest), without vestige of either premise or conclusion in the realm of science. Teilhard's beliefs as to the course and the causes of evolution are not scientifically acceptable, because they are not in truth based on scientific premises and because to the moderate extent that they are subject to scientific tests they fail those tests. Teilhard's mystic vision is not thereby invalidated, because it does not in truth derive from his beliefs on evolution—quite the contrary. There is no possible way of validating or of testing Teilhard's mystic vision of Omega. Any assurance about it must itself be an unsupported act of mystic faith.

Three great men and great souls, and all have flatly failed in their quest. It is unlikely that others can succeed where they did not, and surely I know of none who has. The attempt to build an evolutionary theology mingling mysticism and science has only tended to vitiate the science. I strongly suspect that it has been equally damaging on the religious side, but here I am less qualified to judge.

Does that mean that religion is simply invalid from a scientific point of view, that the conflict is insoluble and one must choose one side or the other? I do not think so. Science can and does invalidate *some* views held to be religious. Whatever else God may be, He is surely consistent with the world of observed phenomena in which we live. A god whose means of creation is not evolution is a false god. Dogma that requires an unchangeable and untestable interpretation of any observable natural phenomena does inevitably conflict with science and cannot possibly coexist rationally with it. That is fundamental, and it is going to keep some religions, or at least some theologies, sadly and unnecessarily at loggerheads with science indefinitely.

Beyond that, religious emotions are very real. They occur to almost everyone and they usually bring pleasure, at the very least, and often deep comfort and an abiding sense of value. Whatever the actual source of these emotions may be, their reality and value exist and may indeed be enhanced by rational consideration of a not necessarily transcendental source. Mysticism quite simply as such, as an unmediated experience *not* rationalized by or related to science, cannot be invalidated (or validated) by science. Teilhard's theory of evolution is flatly untenable, but his vision may have—indeed evidently does have—value for many as it certainly did for him. I have had no vision and I am incapable of faith in the visions of others. That way of religion is closed to me, but that gives me neither right nor reason to deny its reality and value for others. I do not consider the vision *true*—that is beside the point, for truth (to me) is approached only in other ways. I deprecate the vision only to the extent that the desire to prove it true has or may impinge on fields inappropriate to it and must, indeed, have distorted the vision itself.

Man is the moral animal. Moral and ethical systems are necessary for normal human functioning and are major adaptive elements in religion. The propensity for developing moral precepts and the disposition to learn them, as well as the precepts themselves, are adaptations acquired in the course of our biological and social evolution. When viewed in this way, rather than as mere edicts from a stern and incomprehensible source, those precepts achieve a higher sanction and become the more impelling.

The rational, that is, the scientific investigation of the universe reveals its marvels as no amount of introspection or revelation could do. We are in ourselves truly fearful and wonderful, and so is a skylark, a buttercup, or indeed a grain of sand. No poet or seer has ever contemplated wonders as deep as those revealed to the scientist. Few can be so dull as not to react to our *material* knowledge of this world with a sense of awe that merits designation as religious.

Finally, not only the vitalists but all of us want, as Needham has put it

> to ask why living beings should exist and should act as they do. Clearly the scientific method can tell us nothing about that. They are what they are because the properties of force and matter are what they are, and at that point scientific thought has to hand the problem over to philosophical and religious thought.

There lie the ultimate mysteries, the ones that science will never solve. It does not lessen our religious awe of them if we question whether theology, too, is not powerless to pierce that ultimate veil.

PART FOUR

Evolution in the Universe

CHAPTER TWELVE

Some Cosmic Aspects of Evolution

WE have learned that we inhabit a tiny mote lost in an incredibly vast ocean of space and moving inexorably down an endless axis of time. To be sure, our insignificance is not to be overemphasized, for in a sense the mind that encompasses is larger than the universe that is encompassed. Yet it is indeed true that our earth and its life are minuscule and brief in comparison with the cosmos around them. It is inevitable that we should wonder whether this life that we see and share is not a local manifestation of a far more widespread phenomenon. Is it not likely that the evolutionary processes are truly cosmic and not merely terrestrial?

That query leads to thoughts about possible evolutionary histories in general, and not only in the particular configurations that have occurred and now occur on earth. We assume (as in Chapter 7) that the immanent features of the universe are everywhere the same. We know that the configurational features are not, and that actual history has as cause the interplay of the universal immanent and the contingent configurational. Much that bears on the cosmic aspects of evolution has been said in previous chapters. It may here, in part, be brought into this different context, and there are other things to add. Such considerations may then become a basis for judging the probabilities of life, perhaps even of humans or something like them, elsewhere in the universe—questions to be considered in turn in the following chapter.

The historical causation of any event depends on an extremely complicated configuration or momentary condition of organized matter and energy and on the whole sequence of prior configurations through enormous lengths of prior time—back perhaps to the very beginning of the universe, if it had a beginning, or forever, if it had none. For this reason, it must be extremely improbable, at least, that any two configurations, either in different places or at different times, should be exactly the same. Historical causation must be very nearly if not absolutely unrepeatable. At present I only mention that conclusion; some of its implications will be discussed later.

The possibilities of history must be greatly diverse, since historical causation is so enormously intricate and varied. Nevertheless the possibilities cannot be infinite; they must have quite definite limitations. As regards the history of organisms, our enquiry here is made difficult and largely speculative by the fact that we have only one objective example. We know only (and that imperfectly) the history of life on this one planet, a history whose strands are so interconnected that they form a single causal sequence. One of the things we most want to know is how much leeway there may have been in that history. How different would the present configuration be if some past configuration had been different in a given way? Would, for instance, man exist now if some Paleozoic fish had happened to turn north instead of south? Or if that particular species of fish had not existed at all? Or had existed a million years earlier or later than it did?

We cannot directly answer such questions on the basis of our single example. The fish did exist when and where it did and it did act as it acted. That cannot be changed now. We would have a better basis for judgment if we knew about the evolution of life on other planets, where there must have been different configurations in the past. Then we could compare the results of different evolutionary causations. We have no such knowledge yet, and in fact one of the reasons for considering these points is to speculate reasonably on what evolution may have produced elsewhere in the universe. The only available approach is to study how evolution operated on earth and what limitations seem to apply to its operations here.

In the first place, the immanent characteristics of the universe certainly must limit the possibilities of organic evolution quite stringently. Their influence extends to every level of organic activity. At the ultimate biochemical level, for instance, the extraordinarily versatile but highly specific and invariant immanent properties of the carbon atom both make life possible as it is and keep it from being anything else. Henderson long ago and numerous others, such as Blum, since have speculated interestingly on the possibility of an organic chemistry based on some element other than carbon. The possibility of something analogous to life on such a basis cannot be denied, but it is clear that such "life" would be very different from life as it exists here on earth and in ourselves. It could not conceivably produce configurations—such as human beings—even remotely similar to those of our carbon-based organisms.

At the higher, organismic level the restrictions of immanent causation are equally pervasive. For example, physically maximal and (to some extent) optimal sizes of land organisms are governed by the mechanical arrangements of skeletal and muscular systems, the strengths of their materials, and the force of gravity. The possible and optimal sizes and shapes of flying animals, although quite diverse, are held within even more rigid limits, especially in the case of birds, by the forces of gravity, again, and the density of the atmosphere at accessible heights. This points up the interesting fact that the immanent laws, just because they themselves are invariant, must impose variant limitations under different environmental configurations. For example, birds like ours could not function effectively, and therefore would not evolve, on planets where gravity was much stronger or weaker or where the atmosphere was much denser or thinner.

Configurational limitations on evolution are especially pervasive and even more intricate and restrictive. The most important of all these restrictions is paradoxically so obvious that it is likely to be overlooked. It is inherent in the fact that evolution is a historical, continuous process and that it excludes creation entirely *de novo*. Everything in that history—not excepting the first living things—has been a relatively minor modification of an immediate predecessor. Every configuration arises immediately from an antecedent configuration. For any one configuration or, say, species of organism there are probably several or even many different possibilities as to what it can become, but the possibilities are minutely few in comparison with the impossibilities. Some Cenozoic anthropoid demonstrably had the possibility of evolving into several different kinds of apes and also alternatively into man, by changes very small compared with those among groups of organisms in general. But it had no chance at all of evolving into a bird, still less into an octopus, a bee, or a rose. Conversely, the bird, the octopus, the bee, or the rose could not possibly evolve into an ape or a man. And more broadly, what could ever later evolve here on earth has always been strictly limited at every moment through the past two billion years or more by what had already evolved.

Limitations of that kind are involved especially in what Rensch has called the "autonomous laws and rules." By that he means limitations imposed by the nature of the individual organism. Again they run the gamut from the molecular, or even atomic, to the organismic level. At the former level, a gene mutation, which is a change of configuration within a molecule, can only be a modification of something that is already there. The number of different mutations of any one gene must always be limited. (The astronomically large numbers of genetic differences among individuals are not established directly by gene mutations but by recombinations and rearrangements of different genes each with comparatively few mutant or allelic forms.)

At the organismic level, major autonomous restrictions are applied by the fact that the organism has to work; it must remain a going concern. Any innovation must, again, depend on the last previous configuration and be closely adapted to it. To give a gross example, an ameba cannot possibly develop eyes, because it does not have the complex nervous system and other anatomical and functional characteristics that make eyes workable. Or, more realistically, a change in the human nervous system would be unworkable, hence virtually impossible in evolution, if it did not precisely provide for innervation of all the essentially functional muscles.

Beyond that, innumerable restrictions are imposed by what Rensch calls "laws of interaction with the environment." Evolutionary changes must not only arise from and be workable with the configuration of the organism itself, they must also be workable, at least, in the larger configuration in which the organism lives. That is only another way of saying that most evolutionary changes are adaptive. For various reasons about which we need not worry here, nonadaptive changes are also possible and adaptive changes do not always occur. The "laws" in this field may have numerous exceptions. For that reason I think it inadvisable to call them laws and would call them rules or simply generalizations. Nevertheless the directive, nonrandom, or antichance factor in evolution, natural selection, always works effectively in the direction of adaptation. Nonrandom changes are almost invariably adaptive and in conjunction with existing environments (external configurations) this does involve many rules or generalizations of evolution that place statistical or probabilistic, if not absolute, limitations on evolutionary possibilities.

At various times Rensch has compiled some sixty such rules, and those are only examples. (In the mentioned paper he explicitly gives thirty-three.) Three examples will suffice here. (Some others were given in Chapter 11.) (1) Allied races of warm-blooded animals tend to be larger in cooler regions. This generalization, widely known as "Bergmann's rule," is very loose. Rensch found up to 40 per cent of exceptions in mammals of the northern continents. (2) In most lines of evolutionary descent among nonflying animals the body size has tended to become larger. That generalization is striking in fossil sequences, but it evidently has also had many exceptions—otherwise there would by now be no animals smaller than the mechanically imposed limits of size, and there are a great many such. (3) Mammals of tropical regions have relatively shorter and less woolly coats than related forms in colder regions. For this rule, so obviously adaptive, Rensch found no exceptions.

From all these, and also some other, considerations it is clear that evolution is by no means an unlimited or undirected process. The extent and rigidity of the limitation are subjects for further inquiry—or speculation.

Causality is usually taken to imply a unique relationship between causes and effects: identical causes are supposed always to produce identical effects, and identical effects are supposed always to have identical causes. That is a reasonable belief and it is in accordance with most, if not all, human experiences and observations. It is not, however, a logically necessary consequence of the postulate of causality. All events would still be materially determined, or caused, if identical causes sometimes produced different effects and identical effects sometimes followed different causes. There are, indeed, some aspects of evolution that could be taken as reflecting both these possibilities. I am not maintaining that anything we know does or could establish the reality of multiple or ambiguous cause–effect relationships. I think that unlikely, but the possibility is not excluded.

We have already suggested that it is extremely improbable that a historical cause in over-all organic evolution on earth ever has been or ever will be exactly repeated. That causation involves the whole, exceedingly complex configuration of our part, at least, of the universe. Since every element in that configuration is subject to change, and most of them are always changing from instant to instant, precise repetition of the whole pattern is virtually impossible. If, as certainly seems to be the case, there never are identical, over-all evolutionary causes in terrestrial evolution, the question whether identical causes could produce different results lacks reality in this context. Nevertheless it is interesting to speculate as to whether the results *could have* been different from what they actually were. Any conclusion will at least affect our philosophical viewpoint, for instance on the problems of fatalism and free will. Moreover two objective possibilities are not excluded: (1) precise repetition of an over-all evolutionary cause might occur on some other planet in the enormously large—perhaps infinite—universe; and (2) events within the over-all course of evolution on earth not only might but also clearly do sometimes involve pertinent partial causes so similar that it is meaningful to enquire whether their results are equally similar.

If that second point were not true, it would be futile to seek evolutionary rules and principles and meaningless to talk about them. Closely related, warm-blooded animals *are* usually larger in cooler climates, and it is logical to assume that there is a causal relationship (not necessarily at all direct) linking environmental temperature and size. The fact that there are up to 40 per cent of exceptions to the rule might be interpreted as meaning that the same cause can produce different effects, or as indicating that climate is only one of several different determinants of size and that in the exceptional cases we have overlooked important parts of the total causation. The latter alternative is so much more probable as to be almost certain. We do know that total causation is not the same in the animals that do and those that do not follow the rule; the species are different, and their natures are part of the causal relationship. Of course finding the actual and immediate causes of the exceptions (or of the rule, for that matter) is a different and more difficult problem.

That similar causes tend, at least, to produce similar effects in evolution is one of the best-established principles in the field. It is abundantly exemplified by the innumerable instances of convergence among different species of organisms. Regardless of differences in ancestry, organisms that live in similar environments and that adopt similar ways of life (these are partial causes) become in some degree similar in function and structure (these are partial effects). Examples are so familiar that I need mention only a couple. The Tasmanian "wolf" and the true wolf are much alike in appearance and habits, but their less similar ancestries have been quite separate for a hundred million years or so. The golden "moles" of Africa and our true moles are even more similar but have converged from ancestries different for tens of millions of years. Dolphins and the extinct ichthyosaurs, separated in time by some tens of millions of years, resemble each other quite closely functionally and externally but had very different immediate ancestors. Their common stock, radically unlike either of them, is now well over two hundred million years in the past.

In these and all other known examples of convergence the result is never an identity. In spite of all similarities, any competent student can distinguish a Tasmanian "wolf" from a true wolf, a golden "mole" from a true mole, or a dolphin from an ichthyosaur at one glance. The effects are similar in some respects, only, and are different in other respects. This can be immediately related to the complex of evolutionary causes, some similar and some different, all configurational. The configurational resemblances are adaptive: similar environments, similar relationships between the organisms and their environments, and consequent similar action of natural selection. The configurational differences are phylogenetic: different characteristics of ancestors and hence of the organisms themselves when the similar causes affected them. There seems to be a close relationship between similarities and differences of causes and similarities and differences of effects. We are not justified by these phenomena in concluding that there is much, if indeed any, leeway or, so to speak, play in the uniqueness of causation.

The apparently always incomplete nature of convergence is one aspect of the principle of the irrevocability of evolution, which may be stated in this way: the influence of past configurations is not wholly lost in the course of evolution; it continues indefinitely as part of total historical causation. Another and more widely known aspect of the principle is the irreversibility of evolution: organisms do not return wholly to a different ancestral condition. At first sight there may seem to be a contradiction or at least a logical difficulty here. If the ancestral condition is itself part of the cause, why should not a replication of that condition be a possible effect? The answer is that later configurations stated in the principle to be different from the ancestry are also part of the cause and preclude subsequent identity of effect.

As Muller pointed out and as has been repeatedly discussed in different contexts (e.g., by Simpson and Dobzhansky), the elementary processes of evolution at the genetical level are all reversible. This is notably true of mutations, at a level where immanent physical laws predominate and make the causal situation comparatively simple. But even at the genic level reversion becomes extremely improbable when recombination is taken into account. The possible number of gene combinations in any one population is enormously large—so large that not all of them could possibly be formed in real individuals, at least before the genes themselves had changed, after which full reversion is absolutely precluded.

Actually the genes are not recombined quite at random. Their individual frequencies and their recombinations are influenced by natural selection. This brings not only the organism's internal configuration but also the whole configuration of surrounding nature into the causal chain. So far as external configurations are similar, this tends to increase the probability of reversion, or more generally of the limited or partial repetition of historical events. The phenomena of convergence show that this does happen within limits. But the complexities of total configurations and the extreme improbability of their duplication do establish limits and make full identity of results even less probable under the influence of selection than by genetical factors alone.

Dobzhansky has demonstrated experimentally that "an elementary evolutionary event is . . . likely to be repeated," such events being, for example, merely changes of genetical frequencies in effectively identical gene pools under controlled selection, but he emphasizes how unlikely repetition is for "an evolutionary history."

It cannot be said that duplication of historical events within any one evolutionary complex is necessarily excluded in principle, but it is so extremely improbable as to be virtually impossible in actuality.

I have emphasized the extreme improbability, if not impossibility, that different evolutionary configurations should ever become identical, or in more general terms that different total causes should ever have identical effects. Next I propose to consider the fact that any one configuration may have varied outcomes.

It is a commonplace of evolution that multiple species arise from a single original species. The splitting up of lines of descent has always been a prominent feature of the evolutionary pattern. It is not known whether life started with a single kind of organism or with more, but it is clear that the number must have been small. It is also obvious from the fossil record that the many kinds of organisms at one time in the last few hundred million years had a smaller number of ancestors at any previous time, and usually a very much smaller number. (That the total number has not changed more radically during this time results from the fact that *most* lines of descent become extinct.)

The phenomena of phyletic splitting pose an apparent problem in the field of causation and may suggest that identical causes do frequently produce different results. It is, however, easy to see that the problem is only apparent, a pseudoproblem, and that the conclusion, while not excluded, does not follow. We postulate a single ancestral species with, at any one time, a single although complex configuration. We have also postulated that immanent causation is ubiquitous and changeless. Thus both configurational and immanent causes are unitary within a given species at a given time. Nevertheless it can happen and has very frequently happened that a multitude of different effects, of later specific configurations, has resulted. That is the pseudoproblem. (The problem of *how* this happens is no pseudoproblem but is the main matter of the enormous subject of speciation.)

The further fact is that the configuration of any species, although in a sense single, is not uniform. There is heterogeneity within the population, in the structure, genetics, and other characteristics of the individuals, also in the environments of its different members and subgroups, and hence also in the interaction between the two, the most important resultant of which is natural selection. There are factors tending to counteract those heterogeneities, and as long as those factors predominate phylogenetic splitting does not in fact occur. The most important of these is interbreeding among different segments of the population. The crucial factor abetting heterogeneity is reduction and eventual elimination of interbreeding. Once that happens, the elements of heterogeneity predominate and divergence of configuration inevitably begins. It is extremely improbable, hardly possible in the fact, that the internal configurations will be quite the same in any two of the now isolated segments. It is at least equally unlikely that the external configurations of their environments will be exactly alike. Mayr, especially, has shown that such segments are almost always in different geographical environments; when they are not, there is apparently an exaggeration of internal differences between groups. Thus in all probability and without known exceptions, the different effects—the distinct species and evolutionary lineages of common origin—do have distinctly different configurational or historical causes.

The million or more species of insects today are all descendants of a single species that lived in the mid-Paleozoic. All the billions of species of organisms that have ever lived were ultimately derived probably from a single kind, almost certainly from not more than a handful. When you contemplate the almost incredible amount of evolutionary divergence, the thought may arise that *every* possibility may have been realized, that every configuration consistent with the immanent laws has indeed been produced, or at any rate will eventually be so if the process continues indefinitely. If that were true, it would be a logical conclusion that all species were inevitable, that man, for instance, would always appear sooner or later wherever life happened to arise and continued to diversify indefinitely, regardless of the exact nature and sequence of environmental configurations.

There are many reasons for rejecting any such conclusion. Extremely numerous as they are, all the forms of life that ever have or ever will exist on earth must constitute a very small fraction of those that *could* have existed under an indefinitely greater number of different environmental histories. We have already noted the fact that even within single natural populations the number of different combinations of genetic factors mathematically possible is usually so great that only a small fraction of them can ever possibly exist in the form of real individuals. Thus even at the basic genetic level and within any one species it is flatly impossible that evolution should produce and exhaust all its potentialities—and hence should have either to stop or to repeat itself.

In the fossil record and the living world, there is plenty of less theoretical evidence that supports the same conclusion. The limitations of convergence, already discussed, have a decided bearing on this point too. For example, both in Australia and in Eurasia there has long been a variety of ecological niches for large herbivores. On both land masses various of these niches have in fact been occupied by animals whose relationships are rather close in comparison with the whole vast array of organisms—all are mammals of the same subclass. Yet the ecologically most nearly similar forms, the most convergent groups in the two regions, are very different in appearance, habits, and structure: mainly kangaroos in Australia and various hoofed mammals such as horses, deer, cattle, sheep, and antelopes in Eurasia.

The same sorts of considerations apply when successive adaptive radiations in similar environments are compared. There is never any really close approach to identity. The dinosaurs, or more broadly all the Mesozoic reptiles, and the Cenozoic mammals provide a striking case in point. Each group did radiate in a great variety of broadly similar environments, including environments virtually identical except for the reptiles and mammals themselves. Of course there was a great deal of functional convergence. Yet the organisms that thus evolved were really very different. For example, one of the relatively close convergences produced among reptiles the horned dinosaurs and among mammals the rhinoceroses, but in spite of functional similarities those animals are distinctly different in every single anatomical feature.

That example suggests another pertinent point: the improbability that *every* ecological niche is occupied at any one time, that all the species that could exist do in fact exist. As Dobzhansky has remarked in reaching the same conclusion from a different approach, it cannot really be proved that what does not exist might exist. Nevertheless it is extremely improbable that *all* environmental opportunities have ever been exploited by organisms. In our example, the ecological opportunities for dinosaurs in the Cretaceous cannot have been much fewer than for mammals in the Tertiary, and yet the earlier dinosaurs were much less varied than the later mammals.

It is unlikely that even the mammals succeeded in occupying all the niches. Adaptive radiations are probably never complete, nor is convergence a complete and universal phenomenon. There are many Eurasian mammals (weasels, for a single example) toward which no Australian marsupial has converged, and Australian mammals (koalas for example) without convergent Eurasian vicars.

From another point of view, failure to occupy all niches would appear to be inevitable. What actually evolves does not depend only on the environmental configurations. It depends equally on the prior existence of organisms that could evolutionarily exploit the environmental niches. Those organic, specific configurations are widely varied, but by no means indefinitely so. Their possibilities are distinctly limited. Thus what we say is possible when we look only at the environment and its ecological niches may be flatly impossible if we look at the whole situation, including the organisms that may or may not be able to enter those niches.

Two other pertinent aspects of relationships between organismic and environmental configurations should be briefly considered: the multiple ways of achieving comparable adaptions, and the possible divergence between apparent environmental causes and organismic effects. (These, too, have been discussed by Dobzhansky in different terms but with essentially the same conclusions.) The first point is implicit in what has already been said about convergence and divergence, but may now be spelled out. There is, in the old saying, more than one way to skin a cat. In the matter of adaptation, there are commonly so many ways that no two evolutionary lineages adopt just the same way—and incidentally some ways may not happen to have been adopted by any existing organisms.

Simple examples are so frequent and familiar that mention of only a few will suggest a host of others. Kangaroos, horses, birds, and snakes, to mention no others, all have occasion to flee from enemies and all do so quite effectively—but how differently! Still more radically, other animals are about equally effective at evading predation but do so without fleeing. They may be prickly or armored, unbearably offensive in taste or smell, or concealingly colored, or may have any number of other defensive adaptations. They may even be defenseless but simply so prolific that it does not matter for the species (which is what counts) if most of them are eaten in youth. Myriad ways exist for serving exactly the same evolutionary end: that enough individuals escape early predation to breed and to perpetuate the species. By now the reason for such multiple solutions of similar evolutionary problems will at once rise to mind: the particular adaptation acquired depends on what was available, on the preceding configuration, which is never quite the same for any two species. I have elsewhere discussed this aspect of evolution under the term of "evolutionary opportunism": making do with what you have. Of course it does not follow that what a species happens to have will suffice for *any* of the alternative solutions to the problem of survival. That is why in fact the vast majority of species have become extinct.

At the genetic level, Dobzhansky has emphasized that the same elementary adaptive changes, such as an increase in body size, may be accomplished by selection of quite different genes. Since genetic configuration is decidedly a part of the historical causation of evolution, this seems at last to be a substantiation of the possibility that different causes can have identical effects. It is such, however, only in so limited a sense or degree that it may not be a real exception to the apparent rule of the uniqueness of causal sequences. The different genetic factors may indeed produce the same immediate and elementary effect in one particular, but either at the same time or ultimately they are practically certain also to produce differences in other respects. Dobzhansky and Pavlovsky have adduced elegant experimental evidence of this fact. In replicate selection experiments with apparently identical initial populations, the outcome was the same if the genetic background was indeed the same in the experimental populations, but it was variable and (from data on hand) unpredictable if the underlying genetic factors were not identical.

That leads into the last topic of this particular part of our enquiry. Although relationship to environment is the key to the most important and systematic evolutionary changes in organisms, this is far from being a simple matter of environment as cause and organic change as result. If you are dealing with a system of one active and one comparatively inert element, there is usually a close correlation in intensity and kind between the action of one and the effect on the other. If, for instance, you beat a piece of iron with a hammer, the effect on the iron will depend directly and simply on how often and how hard you hit it. But if the element acted on is itself an active reagent, cause and effect may be startlingly disparate. If, for instance, you beat not iron but fulminate with a hammer, there will be an explosion more energetic than your blow, and moreover with no correlation between its energy and that of your blow.

Now of course individual organisms are highly reactive systems in themselves, and reproducing populations are so in even more energetic and complex ways. An environmental change will indeed cause an effect in the population, but, within limits that may be very broad, it will not in itself determine either the nature or the extent of the resultant reaction. In Rensch's example, among certain species of evolving mammals some reacted to northern environments by becoming larger, and in varying degrees even in the same environment, while others reacted by becoming smaller, also in varying degrees. This comes back by another route to the conclusion that the organisms themselves, the genetical and other factors of the configurations of their populations, must be counted among the causes of evolutionary effects on them. That can now be related to the further conclusion that even slight changes in total configuration, or in the environmental parts of configurations, are almost certain to have permanent and far-reaching effects on later evolution and that the magnitude of the effects may be out of all proportion to the initiating cause.

I have now, I believe, laid the groundwork for consideration of some literally cosmic questions: whether life exists elsewhere in the universe; whether there are other manlike beings in the cosmos; whether we are likely to communicate with them; what our own evolutionary future may be. Those questions are the topics of the next two chapters, the last of this book. It occurs to me here, however, that if we are going to consider those problems in any reasonable way we have to make a possibly unexpected assumption to begin with: we must stipulate that the universe we are talking about is finite.

I do not have the slightest idea whether in real fact the universe is finite, or indeed whether the expression "real fact" makes any sense in this connection. Qualified cosmologists (I am not one of them) seem to disagree on that point. Apparently no one is sure, and probably no one ever will be. If, however, the universe were postulated to be infinite, some rather striking consequences would seem to be logically inescapable.

If the universe is not merely vast, which it plainly is, but literally infinite, then there are not a hundred million or a billion billion habitable planets or any enumerable amount of them but an infinite number. Now, in infinity everything that is physically possible, no matter how improbable, must exist and in fact it must exist an infinite number of times. What actually exists here on earth is obviously physically possible. Therefore, if the universe is infinite, somewhere there is another reader exactly like you reading exactly what I have written here. In fact this scene must be repeated an infinite number of times. There must also be an infinite number of readers just slightly different from you reading an essay slightly different from this one. And so on! In that case of course life like ours and men like us do exist on other worlds, on an endless number of them.

I am naive enough to be completely awed by that consequence of a possible infinity. Yet it has no particular bearing on the inquiry in the following chapters. The parts of the universe, if any, that we cannot observe and with which we could not communicate by any means do not "really" exist as far as we are concerned. The parts that we can observe and with which we could (in principle) communicate are by that very fact finite. It would make no sense for us to consider any but this finite universe or, mayhap, finite segment of an infinite universe. A finite segment of space, no matter how large, is infinitesimal in comparison with infinity. The conclusion of endless duplications of life does not at all apply to this infinitesimally finite universe of our discourse. In *it,* as concluded in the next chapter, we are not about to talk to humanoids on other planets.

CHAPTER THIRTEEN

The Nonprevalence of Humanoids

THE possibility that life exists elsewhere than on earth has excited human imagination since antiquity. In our own days it has become the principal basis for a whole school of writing: science fiction, which remains mere entertainment even though some of its devotees do make an unjustified claim that it should be taken more seriously. There has also long been discussion that was scientific, at least in the sense that it was by professional scientists who did not intend to write fiction. Even in the nineteenth century there was serious, if not invariably sober, discussion of the view that life exists not only elsewhere but even everywhere in the cosmos.

There is, then, nothing new in the fact that this subject is being widely discussed and publicized. What is new is that the usual speculation and philosophizing are now accompanied by extensive (and incidentally expensive) research programs, by concrete plans for exploration, and by development of pertinent instrumentation. Although the interested scientists have by no means stopped talking, they are now, and for the first time in history, also acting. Our major space agency, NASA, has a "space bioscience" program. Biologists meeting under the auspices of the National Academy of Sciences have agreed that their "first and . . . foremost [task in space science] is the search for extraterrestrial life" (Hess *et al.*, 1962). The existence of this movement is as familiar to the reader of the newspapers as to those of technical publications. There is even increasing recognition of a new science of extraterrestrial life, sometimes called *exobiology*—a curious development in view of the fact that this "science" has yet to demonstrate that its subject matter exists!

Another curious fact is that a large proportion of those now discussing this biological subject are not biologists. Even when biochemists and biophysicists are involved, the accent is usually on chemistry and physics and not on biology, strictly speaking. It would seem obvious that organic evolution has a crucial bearing on the subject, which is essentially a problem in evolutionary systematics. Surely, then, it is odd that evolutionary biologists and systematists have rarely been consulted and have volunteered little to the discussion. A possible reason for this blatant omission was suggested long ago by an evolutionary systematist, W. D. Matthew, who wrote that, "[Physical scientists] are accustomed to hold a more receptive attitude . . . toward hypotheses that can not be definitely disproved . . . [while] the [evolutionary and systematic] biologist . . . is compelled . . . to leave out of consideration all factors that have not something in the way of positive evidence for their existence."

Matthew also remarked that, "To admit the probability of extra-mundane life opens the way to all sorts of fascinating speculation in which a man of imaginative temperament may revel free from the checks and barriers of earthly realities." Both of his points are illustrated delightfully and without conscious humor by a contemporary leader in exobiology who wrote in 1962, "We do not really know [what the atmosphere of Venus is like], and we are thus not severely limited in our conclusions"! (exclamation point mine).

As an evolutionary biologist and systematist, I believe that we should make ourselves heard in this field. Since part of our role must be to point out "the checks and barriers of earthly realities," we may at times seem merely to be spoilsports, but we do have other contributions as well.

Exobiology has three major questions: "'What kind of life?" "Where?" "How may it have evolved?" Each question in turn involves two complex, distinct fields of inquiry. Confusion of these fields frequently distorts judgment and confuses argument.

The alternative fields as to the kind of life are "life as we know it" and "life as we do not know it." Life as we know it obviously cannot be confined in this context to actual terrestrial species, but implies only a more general similarity. It must, at least, involve a carbon chemistry reacting in aqueous media and with such fundamental organic compounds as amino acids, carbohydrates, purine-pyrimidine bases, fatty acids, and others. It must almost certainly also involve the combination and polymerization of those or similar fundamental molecules into such larger molecules or macromolecules as proteins, polysaccharides, nucleic acids, and lipids. Life as we do not know it might be based on some multivalent element other than carbon, on some medium (perhaps even solid or gaseous) other than liquid water, and then necessarily on quite different kinds of compounds.

If we did encounter such systems or organisms, we might well fail to recognize them as living or might have to revise our conception of what life is. Here on earth, in spite of a border zone between, and enormous diversities within, each realm, we can recognize two kinds of configurations of matter, one living and one not. (Under "configuration" I mean to include not only chemical composition but also organization or anatomy in the fullest sense and energy states and transactions.) "Life as we do not know it," if recognized at all, might have to be recognized as a third fundamental kind of configuration and not, strictly speaking, as life. There has been considerable speculation along such lines, some of it diverting in a science-fictional sort of way. Yet there is not a scrap of evidence that "life as we do not know it" actually exists or even that it *could* exist—evidence, for example, in the form of detailed specifications for a natural system that might exhibit attributes of life without the basis of life as we do know it. (Computers and other artifacts that mimic some features of the life of their makers are not really pertinent to this question.) Here, at least, further consideration will be given only to life as we know it, to the minimal extent of depending on similar biophysical and biochemical substrates.

The dichotomy in discussing the "Where?" of possible extraterrestrial life is between our own solar system and presumed similar planetary systems anywhere else in the universe. Much has been learned over the years about the planets of our system by earth-based astronomical methods. Recently rocketry and telemetry have given us closer looks at the moon and at Venus and promise to give us many additional facts. Human visits to the moon and the closer planets, at least, make no evident further demands on our theoretical knowledge and require only a reasonable extrapolation of our technical potentialities into the near future. Here, then, we have actual observational data to work with, and the promise of many more.

Not so for any planetary systems that may exist outside our own. Statements in both the scientific and the popular literature that there are millions of such systems suitable for life and probably inhabited may give the impression that we know that they do exist. In fact we know no such thing in any way acceptable as sober science. There are no direct observational data whatever. It is inherent in any acceptable definition of science that statements that cannot be checked by observation are not really *about* anything—or at the very least they are not science. As long as we do not confuse what we are saying with reality, there is no reason why we should not discuss what we hope or expect to observe, but it is all too easy to take conjecture and extrapolation too seriously. It is not impossible that our descendants may some day make pertinent direct observations on other planetary systems, but that is far beyond our present capabilities or any reasonable extrapolation from them. With our present techniques, the only way we could obtain direct knowledge of life outside our solar system would be by receiving signals from someone or something out there. That point is involved in the third question, the directly evolutionary one, and its two major fields of enquiry: the origin of life and its subsequent history. Here is my main topic, to which I will return at length.

First it is necessary to refer briefly to the environmental conditions and possible evidence of life on the only planets for which we have any actual data, the planets of our own solar system. Apart from a few eccentrics, astronomers have long since agreed that life as we know it is now quite impossible on any extraterrestrial body in our solar system except Venus and Mars. (See, e.g., the book by Jackson and Moore cited in the notes to this chapter.) Opinion regarding Venus has been divided, but telemetry from the recent Venus probe seems to confirm beyond doubt the previous view that Venus is far too hot for life as we know it (Barath *et al.*). Although somewhat equivocal, such evidence as we have on the composition of the Venusian atmosphere also seems to be unfavorable on balance (see, e.g., Sagan). It would appear, then, that Venus can now be ruled out as a possible abode of recognizable life.

The evidence for Mars is also highly equivocal, but it does not at present entirely exclude the possibility of life there. Temperatures are rigorous and there is little or no free oxygen. Obviously neither man nor any of our familiar animals and plants could possibly live in the open on Mars. Simple microorganisms have, however, been grown in conditions possibly similar to those that just might exist on Mars (Hawrylewicz, Gowdy, and Ehrlich). This possibility depends in part on the usual belief that the so-called ice caps of Mars are indeed composed of water and that the atmosphere is mainly nitrogen with some carbon dioxide. Both beliefs have been authoritatively challenged by Kiess, Karrer, and Kiess, who maintain that the caps are N_2O_4. That and the accompanying concentrations of oxides of nitrogen in the atmosphere would make Mars lethal to life as we know it. In any case, there is increasing doubt that enough water exists on Mars to sustain any form of life.

Direct evidence for life on Mars has also been claimed. The old idea that the so-called canals of Mars were made by intelligent beings no longer merits sober consideration. It is, however, well known that there are dark areas on Mars that show seasonal changes in position and in apparent color. It has been claimed repeatedly that these areas must be covered with some form of plant life, and that idea received significant support when it was discovered that their

infrared spectrum has a band similar to that of some organic compounds (Sinton). However, similar absorption can also be caused by
oxides of nitrogen and by a variety of inorganic carbonates (partly
unpublished work cited by Calvin). The question remains open, and
plans to make direct observations by space probe are going forward
(see, e.g., Levin *et al.*). These plans depend on the further doubtful
proposition that there may be microorganisms on Mars that can be
grown by the same methods used here to grow microorganisms in
laboratories.

The only other direct evidence for extraterrestrial life worthy
of serious consideration is derived from meteorites. It has been
claimed that some of these contain hydrocarbons of organic origin
and even actual fossils of microorganisms (see the articles by Nagy,
Meinschein, and Hennessy; and by Nagy, Claus, and Hennessy). If
confirmed, these observations would indicate that life (now extinct)
had occurred on a planet of our system that has since been disrupted.
However, further investigation strongly suggests that the materials
observed are in part inorganic and in the remaining part terrestrial
contaminants (Anders and Fitch). The most favorable possible verdict is "Not proven."

There is, then, no clear evidence of life anywhere else in our
solar system. Wishful thinking, to which scientists are not immune,
has obviously played a part here. The possibility is not excluded,
but on what real evidence we have the chance of finding life on other
planets of our system is slim.

It bears repeating that there are no observational data whatever
on the existence, still less on the possible environmental conditions,
of planets suitable for life outside our solar system. Any judgment
on this subject depends on extrapolations from what we know of
the earth and its life and from astronomical data that do not include
direct observation. There are, indeed, considerable grounds for such
extrapolations, but they still contain a large subjective element and
have a strong tendency to go over into sheer fantasy.

There are four successive probabilities to be judged: the probability that suitable planets do exist; the probability that life has
arisen on them; the probability that such life has evolved in a pre-

dictable way; and the probability that such evolution would lead eventually to humanoids (as defined in the next paragraph). The thesis I shall now develop, admittedly subjective and speculative but extrapolated from evidence, is that the first probability is fair, the second far lower but appreciable, the third exceedingly small, and the fourth almost negligible. Each of these probabilities depends on that preceding it, so that they must be multiplied together to obtain the over-all probability of the final event, the emergence of humanoids. The product of these probabilities, each a fraction, is probably not significantly greater than zero.

(Before proceeding, I should define "humanoid" for those not as addicted as I am to science fiction. A humanoid, in science-fiction terminology adaptable to the present also somewhat fanciful subject, is a natural, living organism with intelligence comparable to man's in quantity and quality, hence with the possibility of rational communication with us. Its anatomy and indeed its means of communication are not defined as identical with ours. An android, on the other hand, is a nonliving machine, servomechanism, or robot constructed in more or less human external shape and capable of performing some manlike actions.)

The first point, as to the existence of earthlike planets, need not detain us long. The astronomers seem to be in complete agreement that planets that are or have been similar to the earth when life arose here probably exist in large numbers (see the works by Hoyle; Shapley; and Jackson and Moore). Indeed the number of stars in the accessible universe (discernible by light or radio telescopy) is so incredibly enormous that even if the chances of any one of them having such a planet were exceedingly small, the probability that *some* of them do would be considerable. As a basis for further consideration, we may, then, reasonably postulate that conditions such as proved propitious to the origin of life on earth may have existed also outside our solar system.

The next question is: How did life arise on earth, and is it probable or perhaps inevitable that it would arise elsewhere under similar conditions? This is largely in the field of the biochemists, and they certainly have not neglected it. The literature is enormous.

Enough of it for our purposes is summarized or cited in the recent works of Oparin, Florkin, Calvin, and Ehrensvärd. There are wide differences of opinion as to the particular course followed, but here again there is near unanimity on the essential points. Virtually all biochemists agree that life on earth arose spontaneously from non-living matter and that it would almost inevitably arise on sufficiently similar young planets elsewhere.

That confidence is based on chemical experience. If atoms of hydrogen and oxygen come together under certain simple and common conditions of energy, they always deterministically combine to form water. Formation of more complex molecules requires correspondingly more complex concatenations of circumstances but is still deterministic in what seems to be a comparatively simple way. That has, indeed, been demonstrated in the laboratory. If energy such as would be available on a young planet is put into a mixture of the simplest possible compounds of hydrogen, oxygen, carbon, and nitrogen, such as also could well occur on a young planet, amino acids and other building blocks of the essential complex organic molecules are formed. The crucial experiment was that of Miller. A large amount of later work, mostly noted in the books cited above, has confirmed and extended those results. The further synthesis of the building blocks into the macromolecules, especially nucleic acids and proteins, essential for life has not yet been accomplished under realistically primitive conditions. Nevertheless it is reasonable to assume that those steps, too, would occur deterministically, inevitably, if given enough time under conditions likely to hold on some primitive planets. It is also clear that there has indeed been enough time, for the earth is now definitely known to be more than three billion years old, and planets still older could well exist in this and other galaxies.

It is still a far cry from the essential preliminary formation of proteins, nucleic acids, and other large organic molecules to their organization into a system alive in the full sense of the word. This is the step, or rather the great series of steps, about which we now know the least even by inference and extrapolation. A fully living system must be capable of energy conversion in such a way as to accumulate negentropy, that is, it must produce a less probable, less random organization of matter and must cause the increase of available energy in the local system rather than the decrease demanded in open systems by the second law of thermodynamics. It must also be capable of storing and replicating information, and the replicated information must eventually enter into the development of a new individual system like that from which it came. The living system must further be enclosed in such a way as to prevent dispersal of the interacting molecular structures and to permit negentropy accumulation. At the same time selective transfer of materials and energy in both directions between organism and environment must be possible. Systems evolving toward life must become cellular individuals bounded by membranes.

The simplest true organisms have all those characteristics and more, but they are very far from being simple in microscopic and submicroscopic organization. Less organized associations of organic macromolecules, such as are seen today in the viruses, cannot perform all those feats on their own and therefore cannot be meaningfully viewed as primitive and true forms of life.

If evolution is to occur and organisms are to progress and diversify, still more is necessary. Living things must be capable of acquiring new information, of alteration in their stored information, and of its combination into new but still integrated genetic systems. Indeed it now seems that these processes, summed up as mutation, recombination, and selection, must already be invoked in order to get from the stage of loose macromolecules to that of true organisms, or cellular systems. There must be some kind of feedback and encoding leading to increased and diversified adaptation of the nascent organisms to the available environments. Basically such adaptation is the ability to reproduce and to maintain or increase continuous populations of individuals by acquiring, converting, and organizing materials and energy available from existing environments. These processes of adaptation in populations are decidedly different in degree from any involved in the prior inorganic synthesis of macromolecules. They also seem to be quite different in kind, but that is partly a matter of definition and is also obscured by the fact that they must have arisen gradually on the basis of properties already present in the inorganic precursors. In any case, something new has definitely been added in these stages of the origin of life. It requires an attitude of hope if not of faith to assume that the acquisition of organic adaptability was deterministic or inevitable to the same degree or even in the same sense in which that was probably true of the preceding, more simply chemical origin of the necessary macromolecules.

By that I do not mean to say that material causality has been left behind or that some mysterious vitalistic element has been breathed into the evolving systems. All must still be proceeding without violation of physical and chemical principles. Those principles must, however, now be acting in different ways because they are involved in holistic, organic, increasingly complex, multimolecular systems that far transcend simple chemical bonding. It is here that one must stop taking for granted the expectations and extrapolations of the chemist and can obtain further enlightenment only from the biologist as such, that is, the student of whole organisms as they exist in reproducing populations and in communities adapted to environments.

Given ample time and rather simple circumstances not likely to be unique in the universe, there does seem to be considerable probability, perhaps even inevitability, in the progression from dissociated atoms to macromolecules. The further organization of those molecules into cellular life would seem, on the face of it, to have a far different, very much lower order of probability. It is not impossible, because we know it did happen at least once. Nevertheless that event is so improbable that even if macromolecules have arisen many times in many places, it would seem that evolution must frequently or usually have ended at that preorganismal stage. Only the astronomical assurance that there may be many millions of earthlike planets permits us to assume that the origin of true, that is of cellular, life may have happened more than once. In the observable universe the lowest recent estimate for earthlike planets by a competent astronomer is, as far as I know, that of Shapley, who considers 100 million a highly conservative figure. On that basis it is reasonable to speculate that life has arisen repeatedly in the universe, even though we do not know and perhaps will never know whether that is a fact.

Here brief consideration may be given to the idea that once life had arisen somewhere, organisms in a state of cryptobiosis (Keilin) might have spread by "cosmozoan" transport from one planet to another. That possibility was especially urged by Arrhenius, following the still earlier, curious speculation of Richter and others that life may be coextensive with the whole cosmos both in space and in time. It now appears extremely improbable but not quite impossible that any organism, even encapsulated and in a cryptobiotic state of entirely suspended metabolism, could survive the radiation hazards in space without artificial shielding (Becquerel). Furthermore, passage from one solar system to another at any speed attainable by natural means (e.g., by the pressure of light) would require vastly more time than any established or probable duration of the cryptobiotic state, which is not known to have lasted longer than about fifty years in microorganisms or about a thousand years in any organisms (Keilin). A conservative conclusion would be that it is extremely improbable, almost to the point of impossibility, that any form of life has ever traveled by natural means from one planetary system to another. Such travel between earth and Mars, within the same planetary system, is still improbable, but the possibility is not absolutely ruled out.

We now turn to the subsequent evolution of postulated life forms once life has appeared on a planet, and we again move to a different order of probability. We have only a single sample on which to base judgment. Paleobiologists have shown us the general course followed by evolution on this planet. Neobiologists have shown in great, although still incomplete, detail the outcome of that process at one point in time, the present. Although these are far from being the only accomplishments of systematists, they are in themselves so important for current problems as to justify intensified research on this enormous subject.

The problem of extrapolating from this unique sample is to decide whether it is inevitable, probable, improbable, or impossible for life of independent origin to have followed a similar or identical course. Opinions have indeed varied from one end to the other of that scale. I believe that a reasonable choice among those opinions is possible, and furthermore that many, even most, of those who have recently considered the subject have made a wrong choice. Review of recent literature on exobiology, almost all of it by physical scientists and biochemists (or molecular biologists), shows that most of them have *assumed,* usually without even raising the question, that once life arose anywhere its subsequent course would be much as it has been on earth. Now, the only really sound basis for such an assumption would be the opinion that the course followed by evolution on earth is its only possible course, that life cannot evolve in any other way. In a review of two books in which that assumption is made, Blum has called this the "deterministic" point of view as contrasted with an "opportunistic" one. The choice of terms is not a happy one, if only because it is demonstrable that evolution fully deterministic in the philosophical sense would not necessarily, indeed would almost surely not, follow similar courses on different planets. Nevertheless, the two schools of thought do exist and what Blum calls the deterministic one is more commonly followed in current exobiological speculations.

There are here underlying problems of philosophy and indeed also theology. Those problems have been discussed in previous chapters (especially 9, 10, and 11) and need little further attention here. The pertinent *scientific* questions are: If the processes of evolution are the same everywhere as they are here on earth, will they elsewhere lead to the same material results, including men or humanoids? Just how inevitable is that outcome?

Those questions can be followed up in two different but related ways. First, we can examine the course of evolution here on earth to see whether in fact it has proceeded as if directed toward a goal or an inevitable outcome. Second, we can investigate the mechanisms or processes of evolution in order to judge whether and under what conditions their outcome was limited to a course eventuating in some kind of humanoid, that is, in ourselves in the terrestrial example. Those approaches have also been discussed in previous pages (especially Chapters 4, 8, and 12) and need only summaries at this point.

The fossil record shows very clearly that there is no central line leading steadily, in a goal-directed way, from a protozoan to man. Instead there has been continual and extremely intricate branching, and whatever course we follow through the branches there are repeated changes both in the rate and in the direction of evolution. Man is the end of one ultimate twig. The housefly, the dog flea, the apple tree, and millions of other kinds of organisms are similarly the ends of others. Moreover, we do not find that life has simply expanded, branching into increasing diversity, until the organisms now living had evolved. On the contrary, the vast majority of earlier forms of life have become extinct without issue. Usually their places in the economy of nature have then been taken by other organisms of quite different origin. In some cases, their places seem simply to have remained empty for shorter or longer periods.

Neither in its over-all pattern nor in its intricate detail can that record be interpreted in any simply finalistic way. If evolution is God's plan of creation—a proposition that a scientist as such should neither affirm nor deny—then God is not a finalist. But this still does not fully answer the particular question we are pursuing here. The whole nonfinalistic pattern *might* have been followed nearly enough on a planet of some other star to produce humanoids there also. We must turn then to the causal elements and limitations inherent in the process for further judgment of the probability of such an outcome.

Each new organism develops in accordance with a figurative message, coded information, received from its one or two parents. Evolution occurs only if there are changes in that information in the course of generations. Such changes in individuals occur for the most part in two ways, although each takes numerous and sometimes complicated forms: mutations, which introduce new elements into the message, and recombinations, which put these elements into new associations and sequences. In a stricter sense mutations are any changes within the code carried by a nucleic acid. Recombinations involve rearrangements of the various code units and particularly new associations of units from different sources. The latter sources of variation are sexual, and sexlike processes occur in even the most primitive living organisms although they have been secondarily lost in a relatively small number of both plants and animals.

In themselves, these processes are not adaptive; they have no direct relevance to fitting organisms into the economy of nature, permitting their survival and further evolution. Since most (but not all) evolutionary changes are adaptive and progressive evolution does occur, these processes alone cannot be the whole story. They are necessary for evolution, but something else must also be involved. There must be some interaction between organisms and environment and from this there must be some kind of feedback into the genetic code. The feedback is by natural selection and it occurs in populations through successive generations, not in individuals in their lifetimes. That is the whole point of natural selection: that it does feed back from environment to genetic code in such a way as to maintain or change the message in adaptive ways. It does this because, by and large, the better adapted organisms have more offspring. The more adaptive genetic messages thus tend to spread through the population in the course of generations. Also, in more complex ways that I need not go into here, new code combinations adaptive for the population as a whole are thus brought into being.

This feedback is basic for our present enquiry because it places definite limitations on the possible course of evolution. We can be quite sure that if the environments of their ancestors had been very different from what they were, the organisms of today would also be very different. It is also clear that evolution must be opportunistic in the sense that it can work only with what is there. Mutations can occur only in quite definite ways depending on the existing nature of the coded message. Recombination can recombine only the code elements that do exist in given organisms. Selection can work only on variations actually present in a population. The cause of evolution thus includes all the genetic, structural, physiological, and behavioral states of populations right back to the origin of life.

Even slight changes in earlier parts of the history would have profound cumulative effects on all descendant organisms through the succeeding millions of generations. In spite of the enormous diversity of life, with many millions of species through the years, it represents only a minute fraction of the possible forms of life. The existing species would surely have been different if the start had been different and if any stage of the histories of organisms and their environments had been different. Thus the existence of our present species depends on a very precise sequence of causative events through some two billion years or more. Man cannot be an exception to this rule. If the causal chain had been different, *Homo sapiens* would not exist. (These causal limitations were discussed in more detail in the preceding chapter.)

Both the course followed by evolution and its processes clearly show that evolution is not repeatable. No species or any larger group has ever evolved, or can ever evolve, twice. Dinosaurs are gone forever. Nothing very like them occurred before them or will occur after them. That is so not only because of the action of selection through long chains of nonrepetitive circumstances, as I have just briefly noted. It is also true because in addition to those adaptive circumstances there is a more or less random element in evolution involved in mutation and recombination, which are stochastic, technically speaking. Repetition is virtually impossible for nonrandom actions of selection on what is there in populations. It becomes still less probable when one considers that duplication of what are, in a manner of speaking, accidents is also required. This essential nonrepeatability of evolution on earth obviously has a decisive bearing on the chances that it has been repeated or closely paralleled on any other planet.

The assumption, so freely made by astronomers, physicists, and some biochemists, that once life gets started anywhere, humanoids will eventually and inevitably appear is plainly false. The chance of duplicating man on any other planet is the same as the chance that the planet and its organisms have had a history identical in all essentials with that of the earth through some billions of years. Let us grant the unsubstantiated claim of millions or billions of possible planetary abodes of life; the chances of such historical duplication are still vanishingly small.

Even if, as I believe, any close approximation of *Homo sapiens* elsewhere in the accessible universe is effectively ruled out, the question is not quite closed. Manlike intelligence is, after all, a marvelous adaptation, especially in its breadth. It has survival value in a wide range of environmental conditions, and therefore, if it became possible at all, might be favored by natural selection even under conditions different from those on earth. There is, to be sure, another serious hitch here. Man may be going to use one wild aspect of his intelligence to wipe himself out. I do not believe that will occur, but no realist can now deny it as a possibility. If it did happen, the adaptiveness of human intelligence would have been short-lived indeed, and the argument from its apparent broad adaptiveness would be negatived.

Apart from that point, is there not some play, so to speak, in the causations of history? Even in planetary histories different from ours might not some quite different and yet comparably intelligent beings—humanoids in a broader sense—have evolved? Obviously these are questions that cannot be answered categorically. I can only express an opinion. Evolution is indeed a deterministic process to a high degree. The factors that have determined the appearance of man have been so extremely special, so very long continued, so incredibly intricate that I have been able hardly to hint at them here. Indeed they are far from all being known, and everything we learn seems to make them even more appallingly unique. If human origins were indeed inevitable under the precise conditions of our actual history, that makes the more nearly impossible such an occurrence anywhere else. I therefore think it extremely unlikely that anything enough like us for real communication of thought exists anywhere in our accessible universe.

"Extremely unlikely" is not "impossible," and those who like to dream may still dream that mankind is not alone in the universe. But here another point comes up to trouble us. What is the nature and value of that dream? Unless we know or can seriously hope to learn in fact of other humanoids, the dream remains a dream, a fantasy, a science-fiction *divertissement,* a poetic consolation with no substance of reality. Suppose the near-impossible were to be true.

What are the chances that we could in fact learn of the existence of extraterrestrial humanoids and eventually communicate with them? With a feeling almost of sorrow, I must conclude that the chances are vanishingly small.

In the present or any foreseeable state of our technology, the only way we could learn of other humanoids would be by their sending us a message or actually visiting us. That requires, in the first instance, that they must have developed manlike technology, which by no means follows automatically from the mere development of intelligence. (They *might* be intelligent enough to use their brains in better ways!) They must also have done so at just the right time, which involves another tricky point. Out of the billions of years of life on earth, there has been only an infinitesimal length of time, some sixty years, since man has been in a position either to send or to receive messages through outer space. How small the chance of coincidence that any other humanoid reached just this stage at just the right time!

Theoretically, the improbability of humanoids becomes a little less if we consider regions farther out in space and therefore include more stars in our calculations. If humanoids were on a planet a million light years away—and that is a very small distance in the vastness of the galaxies—a message to reach us now would have had to be sent precisely a million years ago. Improbability piled on improbability approaches impossibility. If again the apparently impossible happened, it would certainly be one of the most exciting events in history, but to what avail? The senders of the message would obviously be dead when we received it; their whole species might well be extinct. If, finally stretching the barest possibility to the utmost, we received a message from the relatively nearby stars, it would take years or more likely generations to send a message and receive a reply. Under those conditions the establishment of useful, intelligible intercommunication would still be impossible.

An actual visit to earth by extraterrestrial humanoids would require a technology extremely far advanced beyond ours. We do not, at present, even know that such a stage of technology is possible. All the difficulties previously noted, and more, here pile up. If such a feat is remotely possible and if humanoids are at all prevalent in the universe—the if's do tend to pile up, too, in this subject!—then one would think that we would have been visited by now. In spite of reports of flying saucers and little green men, which belong only in science fiction, the fact is that none have visited us. That would seem, indeed, a logical added reason to believe that humanoids are, to say the least, nonprevalent.

I cannot share the euphoria current among so many, even among certain biologists (some of them now ex-biologists converted to exobiologists). The reasons for my pessimism are given here only in barest suggestion. They will not, I know, convince all or indeed many. There are too many emotional factors and, to put it bluntly, selfish interests opposed to these conclusions. In fact I myself would like to be proved wrong, but a rational view of the evidence seems now to make the following conclusions logically inescapable:

1. There are certainly no humanoids elsewhere in our solar system.

2. There is probably no extraterrestrial life in our solar system, but the possibility is not wholly excluded as regards Mars.

3. There probably are forms of life on other planetary systems somewhere in the universe, but if so it is unlikely that we can learn anything whatever about them, even as to the bare fact of their real existence.

4. It is extremely improbable that such forms of life include humanoids, and apparently as near impossible as does not matter that we could ever communicate with them in a meaningful and useful way if they did exist.

I shall close this chapter with a plea. We are now spending billions of dollars a year and an enormously disproportionate part of our badly needed engineering and scientific manpower on space programs. The prospective discovery of extraterrestrial life is advanced as one of the major reasons, or excuses, for this. Let us face the fact that this is a gamble at the most adverse odds in history. Then if we want to go on gambling, we will at least recognize that what we are doing resembles a wild spree more than a sober scientific program.

To some it seems that the reward could be so great that facing any odds whatever is justified. The biological reward, if any, would be a little more knowledge of life. But we already have life, known, real, and present right here in ourselves and all around us. We are only beginning to understand it. We can learn more from it than from any number of hypothetical Martian microbes. We can, indeed, learn more about possible extraterrestrial life by studying the systematics and evolution of earthly organisms. Knowledge from enlarged programs in those fields is not a gamble because profit is sure.

My plea then is simply this: that we invest just a bit more of our money and manpower, say one-tenth of that now being gambled on the expanding space program, for this sure profit.

CHAPTER FOURTEEN

Man's Evolutionary Future

PREDICTIONS about man's evolutionary future have to be based on assumptions as to how evolution works, and the assumptions made are often wrong.

There is the assumption of orthogenesis, that evolution inherently tends for some mysterious reason to keep on along the same lines. Man's brain increased in size over that of the monkeys, so orthogenetic evolution would keep on increasing the brain until we were fatally top-heavy. But evolution is not really orthogenetic. Trends do not keep on indefinitely but level off, change direction, or even become reversed. Valid predictions cannot be made by extrapolating a past trend into the future. As for man's brain, there is no evidence that it is now increasing in relative size. After the probably rather rapid increase involved in the rise of *Homo sapiens,* brain evolution leveled off and has been practically static ever since.

Another common assumption is that evolution proceeds in abrupt steps by mutations, which set up a new kind of animal or man forthwith. In much of the popular literature mutations are considered as especially likely to give us mysterious new mental powers. But mutations never worked that way in the past, so why should they do so in the future? We have no biological senses that other animals lack; on the contrary, we lack some that others have. Our not wholly new special powers are for tool-using and symbolization, but those certainly accreted gradually and not by single, great mutations. Most mutations are harmful. The few that may become useful are, as a rule, slowly adjusted into the existing genetic system of a species.

Still another of the false assumptions, not the last that exists but the last I shall specify, is that natural selection automatically favors the "fit" in the sense of the strong, the healthy, even the brave and the handsome. Hence come predictions that man will go on getting bigger, healthier, braver, and better looking, and generally improving the breed. Here there is a dangerous half-truth, for natural selection can, indeed, act in that way. However, it does not necessarily do so, and there is good reason to think that it may not be doing so in man.

The present consensus is that natural selection is the usual directive force in evolution. If we knew that natural selection is now effective in man, if we knew its direction, and if we knew that it would continue in that direction, then we could predict much if not all of man's genetical future.

What natural selection favors is simply the genetic characteristics of the parents who have more children. If genetically red-haired parents have, on an average, a larger proportion of children than blondes or brunettes, then evolution will be in the direction of red hair. If genetically left-handed parents have more children, evolution will be toward left-handedness. The characteristics themselves do not directly matter at all. All that matters is who leaves more descendants over the generations. Natural selection favors fitness only if you define fitness as leaving more descendants. In fact geneticists do define it that way, which may be confusing to others. To a geneticist fitness has nothing to do with health, strength, good looks, or anything but effectiveness in breeding.

We must simplify, but it would grossly oversimplify if natural selection were presented as the only important element in evolution. Selection cannot work unless genetic differences among individuals actually exist. Those differences must originally have arisen as mutations of one sort or another. It is further necessary to recognize that selection is seen as an average or statistical effect over a number of generations and that at the same time there are fluctuations in the genetics of populations that seem to be random. As regards particular populations, we should also note that they may be changed quite radically by the exchange of individuals, each with their genetic peculiarities, with other populations. And that brings up the further point that selection occurs not only within but also between populations. Just as some individuals in populations may outbreed others and so influence the course of evolution, so some whole populations may outbreed others and perhaps influence the course of the whole species still more radically.

Natural selection itself becomes extremely intricate when we examine it closely, and it operates in conjunction with numerous other complex factors. It remains true that any long term consistency in the direction of evolution normally has natural selection as the main controlling influence.

Now, is natural selection effective in modern, civilized man? The answer is certainly yes, but not in the way that might be expected. It was perhaps true of some primitive communities that practically every woman who reached child-bearing age did in fact bear children and that all continued to do so at about the same rate until they died or became too old. Little or no natural selection is involved in that situation. But a large proportion of the children died before they reached child-bearing age in their turn. If early death was at all correlated with genetic factors (and it was), that did involve natural selection: selection by mortality, for obviously differential reproduction favored those who lived long enough to reproduce. Such selection must have favored genetical soundness of constitution and resistance to disease. It almost certainly favored intelligence also, for more intelligent parents are likely to give children better care, more intelligent children are more likely to escape dangers, more intelligent families and tribes are likely to cooperate better in obtaining food and shelter for survival, and so on.

That is the usual picture of natural selection improving the breed, and by and large it was probably the correct picture throughout man's evolution until quite recently. Now it is no longer wholly valid in an increasing number of societies. Mortality before child-bearing is not yet negligible, but in the more progressive countries it is comparatively small. There most of the children born live long enough to have children of their own. Children with really gross genetic defects, such as hemophilia or genetic idiocy, still are likely to die young or if they do survive are unlikely to have children of their own. But the far more abundant moderately disadvantageous genetic defects are no longer subject to strong adverse selection, if any. Medical care pulls their possessors through, and these commonly do pass on their characteristics to descendants.

Those well-known facts have led many to conclude, first, that natural selection is ineffective in civilized man and, second, that the result is unmitigatedly bad. The first conclusion is false and the second is at least debatable.

The fact is that the great decrease in early mortality has not led to a relaxing of natural selection but only to a shift of emphasis from one aspect of selection to another. Selective mortality has been largely replaced by selective fertility. In most countries today where early death rates have been radically decreased, the distribution of births has become still more radically unequal. The effect can be measured by the proportions of parents and offspring in successive generations. If, for instance, any random 50 per cent of the parents have 50 per cent of the children, then there can be no selection by differential birth rate or fertility. If, on the other hand, 50 per cent of the parents have, say, 80 per cent of the children, then there is strong, potential selection. The available figures vary considerably and all are evidently incomplete approximations, but they leave no doubt that in most modern civilized communities a comparatively small proportion of parents produces the majority of the next generation. There is thus a high potential for selection. Some students think that selection of this sort in civilized man is markedly more intense than over-all natural selection in primitive man or in most species of other animals.

There is no question that human groups with higher birth rates differ from those with lower rates. In the United States, for instance, it is well known that for a long time parents of low economic and educational status have had more children than those of high status. The difference has recently tended to decrease, but it still exists. Still, no evolutionary tendency can be inferred unless the groups differ genetically. That is the critical point, on which there is much debate, often highly emotional, with few really solid facts.

We do not know the present direction of human selection. Everything said by some can be and is denied by others, if only on the grounds of being unproved. Still there are some broad hints in what data are available. One approach is the empirical one of trying to determine visible changes actually occurring in human populations. Few have been sufficiently substantiated, and of those few perhaps none is universal. The clearest in some populations is an average increase in stature, a change so marked in the United States that it is quite evident to anyone now middle-aged. Even with this change that is known to be in process, we cannot say to what extent, if any, it is genetic, hence really evolutionary, or what the real selective factors—again, if any—are. A more hopeful approach, even though the hope has not yet been fulfilled, is to try to determine genetic characteristics of groups that do now have significantly higher or lower birth rates and that therefore are now favored or opposed by natural selection.

Probably we would all agree that the most distinctive human characteristic now and the one most important for our future is intelligence. Unfortunately at the outset there is the problem of valid comparative measurements of intelligence and even of whether there is a single entity properly specified as intelligence rather than a galaxy of special abilities on which, again, there is no general agreement. Without getting lost in technical details, I think we can take it that there are mental abilities that vary and that can on an average be legitimately compared by means of a number of psychological tests, even though any single comparison may be open to debate. At least there is no doubt that some people are more intelligent than others, even when they have had equal educational and cultural opportunities. Next is the question whether these differences have a genetic basis. The answer is of course disputed, and yet in a general way it does seem reasonably established that intelligence depends in considerable part on heredity. The pertinent data include studies of identical twins, of foster children as compared with foster and biological parents, and others. Effective intelligence is certainly influenced by environment, as most characteristics are, but it has limits set by heredity.

There is thus reason to believe that a population will evolve in the direction of more intelligence if more intelligent parents have more children and in the opposite direction if less intelligent parents have more children. And now comes the crucial step; it is possible if not probable that on an average each new generation nowadays is mostly derived from the less intelligent members of the last generation. Although they are of course disputed, the data are strongly suggestive for the United States, Great Britain, and a few other countries. For large segments of the world's population there simply are no data either way. I know of no evidence for any considerable population that selection is favoring the more intelligent.

It is an unpleasant conclusion that mankind as a whole or at least a considerable segment of it may be evolving in the direction of less intelligence. Many, including some scientists, have indignantly rejected that conclusion, but the grounds for rejection are usually that there is no real proof, that the situation is very complicated, and that there are other possibilities. That is all true, but surely the proper procedure is not to reject what evidence we do have but to seek impartially for more and better evidence. We may not like the truth, but we had better try to learn it for our own good.

The precise genetic basis of intelligence is not known, but it is beyond much doubt that intelligence is influenced by a large number of interacting genes, perhaps by virtually the whole genetic system acting as a complex unit. That seems to be true for most of the characteristics that will really matter in the evolution of mankind, those involved in general and special capabilities and in personality, to the extent that heredity does affect such characteristics. Some other features are largely or wholly determined by a small number of genes. That is true of skin color and many of what may be called "recognition marks"—shape of nose or ears, hair color, and the like

—characteristics that in themselves are really of no importance in any broad view of man's evolutionary destiny.

Most obvious are single mutant genes that are not involved in what we usually think of as normal variation but that produce marked peculiarities, frequently gross defects. Such genes are most noticeable in their effect and are comparatively easy to study, so that we know more about them than about other and really much more important aspects of human genetics. Everyone knows some examples: hemophilia, certain types of dwarfism and idiocy, abnormal short-fingeredness, albinism, and others. Some characteristics so nearly harmless and so common as hardly to be considered abnormal also belong in this group: some kinds of baldness, color blindness, etc.

Most current genetic counseling necessarily concentrates on these fairly clear-cut cases, as did early eugenic programs and as now do also most discussions of genetic effects of radiation. Certainly the load of undesirable mutant genes present in all human populations causes a great deal of individual suffering and places a burden on society. There can be no question that all practicable and ethical steps to reduce that load should be taken. Nevertheless it is my personal opinion that such mutant genes have little importance for the evolution (good or bad) of humanity as a whole.

There are several reasons for that opinion. First, on the record it is unlikely that incidence of single, gross mutations has ever controlled a sustained direction of evolution. Such effects have been postulated, but the actual data are in all cases as well or better explained by other theories. A certain amount of mutation (whether good or bad in itself) is essential for long continued evolution, but its final result depends primarily on natural selection. Selection against uncompensated gross or crippling mutations is so strong that it is almost inconceivable that they can ever have come to characterize any considerable part of a natural population.

In human populations recognizably "bad" genes with gross effects are also subject to natural selection. Amaurotic idiots almost never have children, and hemophiliacs have, on an average, fewer than the populace as a whole. Selection cannot wholly eliminate these defects, any more than eugenic sterilization can, but it keeps them down to a tolerable level—not tolerable from the point of view of the afflicted individuals but tolerable in the sense that they do not appreciably affect the evolution of the population.

A good beginning has been made in counteracting the effects of identifiable "bad" genes, and much greater progress in the future is certain if civilization endures. Here we turn to the argument that these effects of medical science are dysgenic in that they tend to shelter the "bad" genes from natural selection and to cause their spread in the population. I think that argument is fallacious, because a gene is no longer "bad" if in fact its bad effects can be entirely overcome by any practicable means. If, for instance, diabetes mellitus (known to involve strong genetic predisposition) can be fully controlled by simple and universally available medication, then the predisposing genes do no harm and their spread in a population would do no harm.

In the course of evolution our ancestry lost the genetic capacity to synthesize certain vitamins. In the absence of adequate supplies of those vitamins from some outside source, that is a wholly "bad" genetic defect which is common to our whole species. Many thousands of prehumans and humans have certainly suffered and died from it. But as long as we have an adequate environmental source of vitamins in usual foods or in medication, no harm at all occurs and we do not even think of this need as a human defect. Medical control of other defects such as susceptibility to diabetes is not different in principle: it is only a means of fitting the environment to existing genetic systems. Unlike selective mortality it also reduces individual suffering and it saves individuals whose other genetic qualities may be highly desirable.

A striking example of the purely relative nature of "goodness" or "badness" in identifiable mutant genes is provided by the gene *Si*. In homozygous *Si Si* individuals it eventuates in sickle-cell anemia, which is always fatal at an early age. But heterozygotes (*Si si*) do not develop sickle-cell anemia and are resistant to an otherwise highly dangerous form of malaria. Homozygotic *si si* individuals are free of sickle-cell anemia but are likely to die of malaria if they live where it occurs. Now if a population has *Si si* individuals, who get neither sickle-cell anemia nor malaria, it will also have considerable proportions of *Si Si* individuals, all of whom die of sickle-cell anemia, and *si si* individuals, many of whom die of malaria if in a malarial region. These unfortunates pay the price for keeping an effective *Si si* population going in a region of malarial infestation. In that environment the gene *Si* is certainly "good" for the population even though "bad" for the *Si Si*'s.

It happens that we have learned how to eliminate malaria, and where this has been done the *Si si* individuals are no longer favored by natural selection. *Si Si* individuals continue to die from anemia. Therefore in this situation natural selection will rapidly decrease *Si,* now wholly "bad," in the population. If, as happens not to be the case, we had learned how to counteract (or "cure") sickle-cell anemia but had not eliminated malaria, then *si* would have become "bad" and selection would have strongly favored *Si* and opposed *si.* If both sickle-cell anemia and malaria were counteracted, then there would be no selection on *Si* or *si* unless they have other effects now unknown.

To this example, which is quite exceptional in its simplicity, must be added an increasing amount of more general evidence that wild populations are normally highly heterozygous and that it is adaptively advantageous for them to be so. Moreover, natural populations are found to have high percentages of chromosomes with disadvantageous (lethal to subvital) genes, even close to 100 per cent. Nevertheless these populations, as such—as evolving groups—seem in the first place to be well adapted and in the second place not to be changing in the direction indicated by the individually disadvantageous genes.

These facts, with other complications that cannot be followed here, add up to the existence of elaborate buffering systems by which mutant single genes, especially those grossly and obviously disadvantageous in themselves, have little or no long-range adverse effects on evolution and may even have eventually beneficial effects. These systems, surely operative in present human populations, practically assure that "bad" genes will either be kept down to a tolerable proportion or will cease to be "bad" through changes in circumstances, which will increasingly come within human control.

The conclusion here is again that natural selection on whole genetic systems is the thing to watch. The far easier checking on gross genetic defects, so valuable in a humanitarian way, has little bearing on man's evolutionary future.

Most considerations of our future dwell on the technological changes in civilization. Certainly those changes, so obvious to us all in our own lifetimes, are likely to go on at even faster rates and certainly they have tremendous importance for possible human futures. I have little to say about them, partly because that is not my subject and partly because I believe that many aspects of technological advance are really superficial and have no great and basic importance for human evolution. It does not matter much to the man who gets hit whether he is killed by a spear or an atom bomb, and it may not make much difference in our eventual evolutionary fate. There does not even seem to be a really tremendous difference between flying to Europe between meals and strolling across a valley to visit a neighboring tribe. But there is some difference, and in this example we approach what really does seem to be important: technological advances do put us in closer touch with more people, and they make it possible for there to be more people.

It is not necessary to document or to emphasize the fact that a real explosion of world population is now in progress. This inordinate and rapid increase in population will certainly influence and be influenced by man's biological evolution, but here it is particularly difficult to make any clear brief statement of the probable outcome. Not enough is known about the unique phenomenon—no other species ever expanded on such a scale.

It is probable that such a phase of tremendous expansion will intensify selection by differential birth rates. It need not involve significant differences in birth rates among genetically distinct groups, but it enhances the possibility and the potential effectiveness of this aspect of natural selection. On a geographic basis, at least, it is already obvious that such differentials exist. For example, present projections indicate higher birth and survival rates for the near future in Latin America than in any other considerable part of the world. Latin Americans probably have no unique genetic character-istics, but their total genetic makeup, their gene pool, certainly is distinctive in its proportions of genetic traits. Relative increase in the percentage of Latin Americans will therefore change the propor-tions of such traits in mankind as a whole and will constitute evolu-tionary change in the human species. Asians and Africans are also likely to increase proportionately while Europeans decrease. These evolutionary changes are bound to be accompanied by political changes, changes in institutions and forms of government, and acute stresses in the whole social fabric.

Since we have at present no biological criterion by which any race or nation can be judged objectively superior or inferior, there is no way to evaluate the possible biological results of this sort of evolution beyond the plain fact that the future racial composition of *Homo sapiens* will certainly be quite different from the past. Human populations, taken as a whole, are all so nearly equal in important capacities that evolution by selection between such popu-lations cannot be expected to be markedly progressive or retrogres-sive. Selection within populations, of the sort already discussed so inconclusively, will probably also be intensified in this situation and is more likely to involve definite progress or degeneration.

A prediction sometimes made is that increased mobility and population density will break down genetic differences between populations and produce a uniform "racial" type all over the world. On past record and present evidence, that is extremely unlikely. In the first place, even if there were such panmixis, this would in-crease individual variation in the world population or any segment of it. That would probably be a good thing and it may be unfortu-nate that it is not likely to happen. Variation is necessary for evolu-tion to occur, and within broad limits the more variation the bet-ter the chance of evolutionary progress. Widespread recombinations of different genetic factors increase the chances of appearance of new and advantageous combinations. Wide mixture decreases in-breeding, which is usually degenerative in a species with as many "bad" genes as man. Moreover, variation is an asset to any society

that includes a wide variety of useful roles: the ideal genetic qualifications for different civilized occupations can be quite different.

But in fact the outcome of one-world–one-race is so unlikely in
any future we can now reasonably expect that I do not have to
expand on that point. All things may be possible, but this one is
extremely improbable for a few thousand years, at least.

Increase in population cannot go on indefinitely. By any possible technology there is finally a limit to the amount of food that
can be produced on earth. There is indeed a fixed and known
limit to the amount of space on earth. Up to now and for a minority
of mankind, technology has kept ahead of population increase, and
this has suggested to some that there is no problem or that it can
be solved by continued technological advance. An evolutionist must
take a more long-range view, and in the long range that solution is
impossible.

Either this population increase will be stopped by some actions
of man himself, or it will end with humanity packed elbow to elbow,
everyone always undernourished and with enormous death rates. If
there are no strong barriers to interbreeding in such a large population, evolution can be expected to be extremely slow and a state of
evolutionary stagnation may be reached. If, however, there are
numerous local populations each tending to marry within the local
group, evolution within single groups can be appreciable and can
eventually affect the whole species as some groups expand at the
expense of others. As to the directions of such evolution, the probabilities are a matter of opinion about what sort of people are most
likely to survive and to have relatively high birth rates in such circumstances. It could hardly be the more intelligent who would
have high birth rates in conditions of maximum crowding and minimum subsistence.

It must by now be evident that there are many possibilities as
to man's evolutionary future and that the one that eventuates will
depend on factors we do not yet fully grasp and on events that have
not yet occurred. The one that seems to me most probable, although
I find it distasteful, is the one just discussed; intense overpopulation followed by evolutionary stagnation or by evolution in new

directions now really unpredictable but likely to be degenerative. At quite the opposite extreme, it is evidently possible that men will wipe out mankind, thus settling all problems of human evolution most decisively. In the event of widespread warfare with nuclear weapons, it seems more probable that comparatively small groups would survive eventually to repopulate the earth. During their expansion, natural selection would almost certainly strongly favor both physical stamina and mental agility. That might result in a great new step forward in human evolution, even though none of us could approve of the means involved. Other features of a possible future species evolved in that way might depend to considerable degree on chance and on random genetic processes while those groups were still small.

It has often been said that man can, in principle, control his own evolutionary destiny. That really awesome fact involves a unique opportunity and a heavy responsibility. As to how mankind will in fact face up to that responsibility and take advantage of that opportunity, I see no reason for despair but a good deal of reason for pessimism. On the scientific side of the problem, we still lack an enormous amount of essential information. The information is obtainable, and there is ample hope that we will obtain it if the objectives are clearly discerned and urgently sought. Even now, we know enough about the central process of past evolution, natural selection, to make a good start at improving the breeds of *Homo sapiens,* as we have in fact used this knowledge to improve breeds of other species.

There is, furthermore, reason to think that we are on the verge of further biological discoveries that could make selection far more effective or could even supplant it with other, faster and surer evolutionary processes. It is probable that the incidence of mutations can be controlled within broad limits: instances are known in which the rate of mutation is itself a genetic factor subject to selection. Control over the direction of mutation, possible now only in a few quite special cases, is another eventual probability. Growing knowledge of the actual chemical nature and structure of genes holds the possibility that genes or in the end even whole genetic systems can be made to order. The guidance of evolution could then be-come a simple matter of following specifications. That possibility is, however, probably in a very distant future, and in the meantime the processes of natural selection are most effective right now and the most likely means of further advance—or retreat.

The most serious question is whether any way to achieve our evolutionary objectives could really be followed as a political rather than scientific problem. Control of human evolution, by any conceivable means, must involve some measure of control over hu-man reproduction both in quantity and in quality. Strict dictation without concern for the consent of the reproducers could achieve that aim. Surely all of us consider that solution not only unethical but also unbearable. But the consent of the reproducers, and their cooperation, will require a degree of comprehension and of altruism that is not now even remotely in sight. It is possible to imagine some first steps, such as social or economic penalties against large families in the population as a whole but rewards for large families in elite, genetically superior groups. The very words "elite" and "superior" arouse reactions that show how far such a first step is from present acceptance.

Such, then, is the human dilemma. Man is a glorious and unique species of animal. The species originated by evolution, it is still actively evolving, and it will continue to evolve. Future evolu-tion could raise man to superb heights as yet hardly glimpsed, but it will not automatically do so. As far as can now be foreseen, evo-lutionary degeneration is at least as likely in our future as is further progress. The only way to ensure a progressive evolutionary future for mankind is for man himself to take a hand in the process. Al-though much further knowledge is needed, it is unquestionably pos-sible for man to guide his own evolution (within limits) along de-sirable lines. But the great weight of the most widespread current beliefs and institutions is against even attempting such guidance. If there is any hope, it is this: that there may be an increasing num-ber of people who face the dilemma squarely and honestly seek a way out.

NOTES AND REFERENCES

Chapter 1 The World into Which Darwin Led Us

THIS CHAPTER is a slightly modified version of a public address given at the Chicago meeting of the American Association for the Advancement of Science on 29 December 1959, one of the activities celebrating the centenary of the publication of *The Origin of Species*. The original version was published in the Association's journal, *Science*, CXXI (1 April 1960), pp. 966–74.

In the present book, the chapter both expresses a theme and introduces numerous topics pursued further in later chapters. Such commentary and reference to other literature as is indicated will be given in connection with that extended discussion. I should, however, refer at once to the invaluable variorum edition of Darwin's *The Origin of Species*, edited by Morse Peckham (Philadelphia: University of Pennsylvania Press, 1959). Darwin greatly modified that book—often for the worse—in the course of the six editions during his lifetime. Only in the variorum can those historically important changes be clearly followed. For straightforward reading, there are numerous reprints currently available, almost all of the sixth edition. The first edition, superior in some ways, is now in process of being reprinted by the Harvard University Press.

The quotation from T. H. Huxley is from *The Life and Letters of Charles Darwin*, edited by Francis Darwin (New York: Basic Books, Inc., 1959).

Chapter 2 One Hundred Years
Without Darwin Are Enough

A SLIGHTLY different version of this chapter was published under the same title in *Teachers College Record*, LX (May 1961), pp. 617–26.

Some further remarks on education and on how I happened to become a student of evolution are given in my chapter on zoology as a career in *Listen to Leaders in Science* (Atlanta: Tupper and Love; New York: Holt, Rinehart, & Winston: scheduled for publication in 1964).

The futility and errors of the Scopes trial are considered at somewhat greater length in my review of *Six Days or Forever* by Ray Ginger. *Nation*, CLXXXVI, No. 19 (10 May 1958), pp. 420–21.

Chapter 3 Three Nineteenth-Century
Approaches to Evolution

THE SUBSTANCE of this chapter was first given as a lecture at Vassar College in 1961, and the written form, almost exactly as it is here, was published in the *American Scholar*, XXX, No. 2 (Spring 1961), pp. 238–49.

The most pertinent works of Lamarck, Darwin, and Butler are sufficiently identified in the text. The variorum *Origin of Species*, mentioned in the notes to Chapter 1, is of course again essential here. There is an English translation of the *Philosophie Zoologique*, but I have preferred working from the original. The two short passages here quoted in English were translated by me. If they seem somewhat turgid and obscure, that is at least in part a faithful reflection of the original. The fact that Lamarck's style was rarely polished and not always lucid was one reason for his not making a deeper impression on cultured French contemporaries.

Guyénot's brief but valuable comments on Lamarckism and Neo-Lamarckism are in *Les Sciences de la Vie aux XVIIe et XVIIIe Siècles* (Paris: Éditions Albin Michel, 1941). An excellent discussion of Lamarck by one of the few who know and care what Lamarck really said is that by Gillispie in *Forerunners of Darwin*, edited by B. Glass, O. Temkin, and W. Straus, Jr. (Baltimore: The Johns Hopkins Press, 1959).

The edition of Darwin's autobiography mentioned in the text is *The*

Autobiography of Charles Darwin, 1809–1882, with Original Omissions Restored, edited with appendix and notes by his granddaughter Nora Barlow. (London: Wm. Collins Sons and Co. Ltd., 1959. An American edition is published by Harcourt, Brace & World.) This book obviously must be read by all Darwin buffs. I reviewed it at length in *Scientific American,* CXCIX (August 1958), pp. 117–22. Glancing over that review again, I note a quotation from Norman Douglas's *South Wind* that is quite pertinent here. Keith, the aging hedonist, is discussing Butler with Denis, the romantic youth. Keith says of Butler: "It was an age of giants— Darwin and the rest of them. Their facts were too much for him; they impinged on some obscure old prejudices of his. They drove him into a clever perversity of humor. . . . Anything to escape from realities—that was his maxim. . . . He personifies the Revolt from Reason. . . . He understood the teachings of the giants . . . but they irked him. To revenge himself he laid penny crackers under their pedestals. His whole intellectual fortune was spent in buying penny crackers."

The literary biographer of Darwin who went so badly astray as to the significance and present status of Darwinian theory is Gertrude Himmelfarb.

The popularity of the catchword "the two cultures" dates from C. P. Snow's *The Two Cultures and the Scientific Revolution* (New York: Cambridge University Press, 1959).

Chapter 4   A Modern Approach to Evolution

THE PLAN of this book here called for a summary of main points in modern evolutionary theory. I had confidently expected to supply that by adaptation of some previous publication, perhaps the chapter *The Study of Evolution* from the symposium *Behavior and Evolution* edited by my wife, Anne Roe, and me. (New Haven: Yale University Press, 1958.) Although I have lifted an idea or two from that chapter, re-examination showed it to be quite inappropriate for present purposes. Indeed I was somewhat shocked to find that no single previous lecture or essay of mine has attempted a modern, over-all view of evolutionary theory, and this chapter has had to be written anew.

So brief a statement on so large a subject of course leaves out innumerable details, subtleties, and exceptions. There is a temptation to supply annotations exceeding the text in length. Without yielding to that temptation, I shall just exemplify the kinds of things that do call for amplification and then cite a few works in which such further information may be found.

Inheritance has here been discussed in terms of nucleic acids, especially of DNA in chromosomes, although it is mentioned that the nucleic acid primarily involved is not quite invariably DNA or without exception in chromosomes. In fact there is always at least one further important factor in heredity. Every organism, whether remaining unicellular throughout life like an ameba or becoming inordinately multicellular like us, starts out as a cell that is *already* highly organized. That organization involves chemical composition, regional differentiation, organelles, cytoplasmic inclusions, and so on. It is essential in determining what the organism is and becomes, and it is derived from the parent. It is therefore a form of heredity. It is, however, part of a parental system that also involves DNA control, and it can evolve little if at all in real independence from DNA evolution. Concentration on the latter is thus justified even though it is not the whole story.

The discussion in the text applies to organisms among which offspring at least occasionally receive genetic material from more than one source, typically from male and female parents but sometimes in other ways. That is true of the great majority of organisms, from simplest to most complex, but there are some that have in all likelihood entirely lost that sexual (or sometimes parasexual) factor in reproduction. In them there is no recombination or gene migration and there are no species in the same sense as in interbreeding organisms. (Whether, as I do maintain, there are species in another sense is a moot question that I shall not debate here.) Evolution in them is essentially a simple interplay of mutation and natural selection, lacking the complexity and flexibility of the more usual and more important processes discussed in the text.

Those uniparental organisms constitute a limited, partial exception to the conclusion that single mutations do not, in themselves, form new species or even constitute significant evolutionary change. In such unusual groups a mutation may eventuate, in a few generations, in a population at least analogous to a species. There is one other kind of known exception. By some variation in genetic processes, it can occasionally happen that an offspring gets a doubled (or otherwise unusual) number of chromosomes. Sometimes this makes interbreeding with normal members of the parental species impossible and yet does not prevent the abnormal offspring from producing fertile descendants of its own. Then a new species may result from a single genetic event, although not strictly speaking a gene mutation. This is fairly common among plants but rarely if ever leads to really marked evolutionary change. It is a very unusual event among animals, probably never occurring in most groups, and still less important over-all than in plants.

As a final example of things left out in the text, natural selection does not necessarily produce evolutionary change. In fact its most frequent and universal role is probably in slowing down or preventing such change. We have noted genetic processes within populations that are not oriented toward adaptation and that would therefore usually tend to disrupt adaptation if they were not checked. It is natural selection, in the form of stabilizing selection, that checks them, that keeps the population from changing in an inadaptive way. An aspect of the process is that extreme variants generally have fewer offspring than more average members of a population. That is simple enough, but in detail the action of stabilizing selection can become very intricate.

The following are a few of the many works helpful in the next step of learning about the mechanisms of evolution:

Dobzhansky, Th. *Genetics and the Origin of Species,* 3rd ed. (New York: Columbia University Press, 1951).

———. *Evolution, Genetics, and Man* (New York: Wiley, 1955).

Grant, V. *The Origin of Adaptations* (New York: Columbia University Press, 1963).

Huxley, J. S. *Evolution: the Modern Synthesis* (New York: Harper, 1943).

Lerner, I. M. *Genetic Homeostasis* (New York: Wiley, 1954).

Mayr, E. *Animal Species and Evolution* (Cambridge: Harvard University Press, 1963).

Moody, P. A. *Introduction to Evolution* (New York: Harper, 1962).

Rensch, B. *Evolution Above the Species Level* (New York: Columbia University Press, 1959).

Roe, A., and G. G. Simpson, eds. *Behavior and Evolution* (New Haven: Yale University Press, 1958).

Schmalhausen, I. I. *Factors of Evolution* (New York: Blakiston, 1949).

Simpson, G. G. *Major Features of Evolution* (New York: Columbia University Press, 1953).

———, C. S. Pittendrigh, and L. H. Tiffany. *Life: An Introduction to Biology* (New York: Harcourt, Brace & World, 1957).

Stebbins, G. L., Jr. *Variation and Evolution in Plants* (New York: Columbia University Press, 1949).

Tax, S., ed. *Evolution after Darwin* (Chicago: University of Chicago Press, 1960).

Chapter 5 Biology and the Nature of Science

I GAVE a lecture on this subject at the dedication of Lapham Hall, University of Wisconsin–Milwaukee, in February 1962, and the text of the lecture was printed for private distribution by the University. An extensively revised version was published in *Science*, CXXXIX, No. 3550 (11 January 1963), pp. 81–88. Except for the first paragraph and a word here and there, the present chapter is the same as the essay in *Science*.

As with Chapter 4, there is so much more to say on the topics of this chapter that full annotation would occupy the rest of the book. With great restraint, I shall expand on only two points in the form of notes taken from the University of Wisconsin private printing:

1. "Material" and "objective" are words often loosely used and hard to define in a precise and useful way. By "material" I mean existent as phenomena in the universe accessible to our (aided or unaided) perception as opposed to the spiritual, the supernatural, or the ineffable. By "objective" I mean phenomena that exist in their own right, independent of what we may think about them. (Whether in fact there are any such phenomena is a point touched on in the essay.) The main test of objectivity is whether observation of the phenomena can, in principle at least, be repeated by others. In this sense, inner thoughts and feelings are not objective and cannot be studied scientifically, but their results may be objective and what is said about them by those who experience them is objective. (That is, the statement itself is objective; its referent may not be.) Clinical psychology scientifically studies these nonobjective entities by such objective manifestations. Even unmanifested thoughts and emotions are material to those who experience and perceive them.

2. Further extension of the argument toward the end of this chapter might lead to the view that sociology, broadly construed, would be the central and unifying science. If scientific endeavor continues long enough and with enough success, I believe that may ultimately be true. As a science, however, sociology is still in far too primitive a state to take on that role. Biology is now mature enough for the role, both through the grasp of its own special subject matter and through the virtual completion of its bridges to the physical sciences. It is largely that last point that justi-

fies what would otherwise have been inexcusable overemphasis on biophysics and biochemistry in recent years. Moreover it is the life sciences as a whole that I see as embodying the essence and goal of science over-all and not particularly some one special life science. In that broad sense sociology, too, is a life science—certainly its subject matter is alive!

The works of Plato, Aristotle, Ptolemy, Kepler, Bacon, and other early worthies mentioned here hardly need citation as being well known or, at least, available in summary in any work on the history of science or of civilization.

More recent studies discussed or mentioned by the authors' names in the text are:

Bridgman, P. *Reflections of a Physicist* (New York: Philosophical Library, 1955).

———. *The Ways Things Are* (Cambridge: Harvard University Press, 1959).

Campbell, N. *What Is Science?* (New York: Dover Publications, 1952).

Conant, J. B. *On Understanding Science* (New Haven: Yale University Press, 1947).

Gillispie, C. *The Edge of Objectivity* (Princeton: Princeton University Press, 1960).

Jeans, J. *The New Background of Science* (New York: Macmillan, 1934).

———. *Physics and Philosophy* (New York: Macmillan, 1943).

Mayr, E. "Cause and effect in biology." *Science,* CXXXIV (1961), pp. 1501–6.

Pearson, K. *The Grammar of Science* (London: W. Scott, 1892).

Pittendrigh, C. S. "Adaptation, natural selection, and behavior," in *Behavior and Evolution,* edited by A. Roe and G. G. Simpson (New Haven: Yale University Press, 1958), pp. 390–416.

Russell, B. *The Scientific Outlook* (Glencoe, Illinois: The Free Press, 1931).

Chapter 6   *The Study of Organisms*

A SLIGHTLY DIFFERENT version of this chapter was published under the title *The Status of the Study of Organisms* in the *American Scientist,* L, No. 2 (March 1962), pp. 36–45.

Much of the thought here (and indeed in other parts of this book as well) is based on the work of my friends Th. Dobzhansky, Ernst Mayr, and

Colin S. Pittendrigh and on conversations with them. They all also read the first draft of this chapter and made suggestions leading to considerable revision.

This brief statement of status within the life sciences could be supported by hundreds of references, but really needs none. I shall give just one because I was myself involved in it and because its very inconclusiveness and, at times, incoherence illustrates all too well the problems of classification and organization in the life sciences and of dialectic among their practitioners: *Concepts of Biology,* a symposium edited by R. W. Gerard. *Behavioral Science,* III, No. 2 (April 1958), pp. 89–215.

Chapter 7   *The Historical Factor in Science*

IN THE SEQUENCE of development of my thought, this subject inevitably followed those of the last two chapters. The factor in science too little envisioned in a physical philosophy and the factor in biology that raises ultimate problems not soluble in test tubes is the historical factor. While I was pondering those points, I was asked by C. W. Albritton, Jr. to contribute on a topic of my own choice to a symposium on the philosophy of geology. Geology is a historical science, too, and I just wrote out my thoughts on the subject in that context. That version has appeared under the title "Historical Science" in the book *The Fabric of Geology,* edited by C. W. Albritton, Jr. (Reading, Mass., Palo Alto, London: Addison-Wesley, 1963), published in celebration of the seventy-fifth anniversary of the Geological Society of America.

The present chapter, somewhat abbreviated and considerably revised from the previous version, returns more to my original context and intention. Its emphasis is less geological and more biological and evolutionary.

The book on the philosophy of geology mentioned above contains more on this subject, particularly the chapter "The Theory of Geology" by David B. Kitts. I think that Kitts has been unduly impressed by the physicists and physical philosophers. His study is nevertheless an excellent synthesis of paleontology, history, and philosophy.

The publications cited in the text of this chapter are as follows:

Bernal, J. D. *The Physical Basis of Life* (London: Routledge & Kegan Paul, 1951).

Braithwaite, R. B. *Scientific Explanation* (New York: Cambridge University Press, 1953).

Conant, J. B. *On Understanding Science* (New Haven: Yale University Press, 1947) .

Farrand, W. R. "Frozen Mammoths." *Science,* CXXXVII (1962), pp. 450–52. [See also the article by Lippman.]

Gillispie, C. C. *Genesis and Geology: A Study in the Relations of Scientific Thought, Natural Theology, and Social Opinion in Great Britain, 1790–1850* (Cambridge: Harvard University Press, 1951).

Hempel, C. G., and P. Oppenheim. "The Logic of Explanation." In *Readings in the Philosophy of Science,* edited by H. Feigle and M. Brodbeck (New York: Appleton-Century-Crofts, 1953), pp. 319–52.

Hobson, E. W. *The Domain of Natural Science* (Aberdeen: The University, 1923).

Lippman, H. E. "Frozen Mammoths." *Science,* CXXXVII (1962), pp. 449–50. [See also the article by Farrand.]

Mayr, E. "Cause and Effect in Biology." *Science,* CXXXIV (1961), pp. 1501–6.

Morgan, C. Lloyd. *Emergent Evolution.* (London: Williams and Norgate, 1923).

Nagel, E. *The Structure of Science: Problems in the Logic of Scientific Explanation* (New York: Harcourt, Brace & World, 1961).

Rensch, B. "The Laws of Evolution." In *Evolution After Darwin,* edited by S. Tax (Chicago: University of Chicago Press, 1960), Vol. I, pp. 95–116.

Scriven, M. "Explanation and Prediction in Evolutionary Theory." *Science,* CXXX (1959), pp. 477–82.

Simpson, G. G. "The History of Life." In *Evolution After Darwin,* edited by S. Tax (Chicago: University of Chicago Press, 1960), Vol. I, pp. 117–80.

Toynbee, A. *A Study of History* (London: Oxford University Press, 1945).

Chapter 8   *The History of Life*

THIS IS a considerably modified version of the last part (pp. 152–77) of a long essay first published under the same title as this chapter in the first volume of *Evolution After Darwin,* edited by S. Tax (Chicago: University of Chicago Press, 1960).

The following is a geologic time scale for reference purposes:

Eras	Periods	Epochs	Approximate date of beginning, in millions of years before the present
CENOZOIC	Quaternary	Recent	
		Pleistocene	2
	Tertiary	Pliocene	10
		Miocene	25
		Oligocene	35
		Eocene	55
		Paleocene	70
MESOZOIC	Cretaceous		135
	Jurassic		180
	Triassic	[Epoch	230
PALEOZOIC	Permian	names	280
	Carboniferous	not in	345
	Devonian	general	405
	Silurian	use]	425
	Ordovician		500
	Cambrian		600
PRECAMBRIAN			More than 3000

Although based on increasingly precise methods of radiometry, the given dates in years still represent orders of magnitude and not accurate measurements of real dates.

Discussion in the text often refers to groups of organisms that are more or less inclusive or distinctive. That is done in terms of the Linnaean hierarchy, which is as follows, in sequence from most to least basically distinct and broadly inclusive groups:

Phylum
  Class
    Order

Family
Genus
Species

More directly descriptive-narrative accounts of the history of life, as well as names and characterizations of the various groups of organisms involved, can be found in books on historical geology. *History of the Earth* by Bernhard Kummel (San Francisco: W. H. Freeman and Co., 1961) is a particularly good one. I have provided a simple discussion of how the fossil record is read and of the recorded forms of life in *Life of the Past* (New Haven: Yale University Press, 1953).

References to Darwin here are to the sixth (1872) edition of *The Origin of Species*. Reference to Lady Barlow's edition of the autobiography is given above, in the notes to Chapter 3.

Adaptive radiation, trends, and numerous other aspects of the pattern of the history of life are discussed, with citations of other literature, in my *Major Features of Evolution* and other works listed in the notes to Chapter 4.

On the multiple parallelism in reptilian-mammalian transitions see: Simpson, G. G. "Evolution of Mesozoic Mammals." *International Colloquium on the Evolution of Lower and Non-specialized Mammals* (Brussels: Kon. Vlaamse Acad. Wetensch, 1961). Part I, pp. 57–95.

The data for Table 4, and other interesting technical information on this subject, are in:

Olson, E. C. "Origin of Mammals Based upon Cranial Morphology of the Therapsid Suborders." *Geol. Soc. Amer., Special Papers* No. 55 (1944).

Chapter 9   Evolutionary Determinism

THIS CHAPTER is taken without essential change from an essay published in the *Scientific Monthly*, LXXI (1950), pp. 262–67, under the title *Evolutionary Determinism and the Fossil Record*. Some of the points made here are also considered in other chapters, written later, but there is no contradiction. This brief essay may here serve as a partial summing up of what has gone before and introduction to the next topic.

Essential references include:

Berg, L. S. *Nomogenesis; or Evolution Determined by Law* (London: Constable, 1926).

Sewertzoff, A. N. *Morphologische Gesetzmässigkeiten der Evolution* (Jena: Gustav Fischer, 1931).

Simpson, G. G. "L'orthogénèse et la théorie synthétique de l'évolution." *Colloques Internationaux du Centre National de la Recherche Scientifique* [Paris], XXI (1950), pp. 123–63.

Vandel, A. 1949. *L'Homme et l'Évolution* (Paris: Gallimard, 1949).

On the various "laws" of evolution, see the article by Rensch, cited in Chapter 12.

Chapter 10   *Plan and Purpose in Nature*

THE SUBSTANCE of this chapter was given as a Vanuxem Lecture at Princeton University on 8 January 1947, and a written version was published the following June in the *Scientific Monthly* (LXIV, pp. 481–95) under the title of *The Problem of Plan and Purpose in Nature*. Revision for republication here has been largely limited to correction of expressions that would be anachronistic at this later date and to elimination of material that would be repetitious in this new context.

The Vanuxem lecture from which this chapter distantly derives has a special retrospective significance for me. It was the first public expression of a growing interest in evolutionary philosophy that was greatly expanded the following year (1948) in the Terry Lectures at Yale University, and then in *The Meaning of Evolution* (New Haven: Yale University Press, 1949). Later studies, including others assembled in the present volume, have carried the interest still further.

Chapter 11   *Evolutionary Theory: The New Mysticism*

THE DISCUSSION of Teilhard is adapted, with considerable modification, from my review of *The Phenomenon of Man* in the *Scientific American* [CCII (April 1960), pp. 201–7]. The rest of the chapter is new.

The works primarily under discussion here are sufficiently identified in the text. The terminal quotation from Joseph Needham is in the volume that he edited, *Science, Religion and Reality* (New York: Braziller, 1955. The text is unchanged from the original issue of 1925, but there is a new introduction by George Sarton). Needham's generally excellent essay includes an example, both sad and funny, of the failures of nerve and logic so likely to occur in this field. Needham most skillfully exposes the fallacies of vitalism and conclusively demonstrates the sterility and invalidity of that philosophy. But he then concludes that mind cannot be explained by a mechanistic [or materialistic, or naturalistic, or nonvitalistic] biology— blithely ignoring the (it would seem) obvious fact that all his arguments against vitalism apply with full force to this conclusion of his. He goes on to say that, "Living matter is . . . the splash made by the entry of mental existences into the sea of inert matter." That would be extraordinarily silly even if it came from an avowed vitalist and mystic.

In fact so many silly things have been said in the name of vitalism that even the vitalists do not like to accept that name. Both Lecomte du Noüy and Sinnott explicitly claim not to be vitalists, but they are, just the same. They both are also quite explicit that living matter involves principles not present in nonliving matter, and that makes them vitalists by definition. I have also labeled Teilhard as a vitalist, but he was that in a different and extraordinary sense. He also believed in a nonmaterial principle, variously labeled as psyche, *conscience* (in French), mind, radial energy, soul, spirit, and the within. This is, under the usual definition, the vital principle basic to philosophical vitalism, but Teilhard believed that it exists in *everything,* nonliving as well as living—indeed he abandoned any fundamental distinction between life and nonlife.

Anyone especially interested in the interplay of science and religion should read the never superseded classic *A History of the Warfare of Science with Theology in Christendom* by A. D. White. Originally published in 1896, it is now available in a two-volume paperback reprint by Dover Publications, New York. Huxley's *Religion Without Revelation,* cited in the text, has a good short bibliography on this and related subjects, although it curiously omits White's book.

Chapter 12   *Some Cosmic Aspects of Evolution*

I GAVE a lecture on the subject of this chapter at Bowling Green State University, Ohio, on 28 February 1960. A revised version of the lecture was later published in *Evolution und Hominisation* (Stuttgart: Gustav Fischer, 1962), edited by G. Kurth, a *Festschrift* to Professor Gerhard Heberer, whom I am happy to greet again in this note. The present chapter embodies much of the text as printed in the Heberer *Festschrift,* but is curtailed and variously modified in other ways.

The following are publications explicitly cited in the text:

Blum, H. F. *Time's Arrow and Evolution* (Princeton: Princeton University Press, 1955).

Dobzhansky, Th. "Evolution and Environment." In *The Evolution of Life,* edited by S. Tax (Chicago: University of Chicago Press, 1960), pp. 403–28.

———— and O. Pavlovsky. "Indeterminate Outcome of Certain Experiments in *Drosophila* Populations." *Evolution,* VII (1953), 198–210.

Henderson, L. J. *The Fitness of the Environment* (New York: Macmillan, 1913).

Mayr, E. *Systematics and the Origin of Species* (New York: Columbia University Press, 1942).

Rensch, B. "The Laws of Evolution." In *The Evolution of Life,* edited by S. Tax (Chicago: University of Chicago Press, 1960), pp. 95–116.

Simpson, G. G. *The Meaning of Evolution* (New Haven: Yale University Press, 1949).

————. *Major Features of Evolution* (New York: Columbia University Press, 1953).

Chapter 13   *The Nonprevalence of Humanoids*

IN THE SPRING of 1963 I gave lectures on this subject (but entitled "Life on Other Worlds") at six member institutions of the University Center of Virginia and at the University of Colorado. The present chapter, not printed before in any version, is based on those lectures but has been extensively revised.

I have stressed that "there are no direct observational data whatever" on any planetary systems but our own. On 19 April 1963 the New York *Times* announced that Dr. van de Kamp of Swarthmore had discovered the third such planetary (or "solar") system. The apparent contradiction is a matter of definition of "direct observation" and "solar system" and really calls for no correction of my text. Three stars are inferred to have *unobserved* dark companions on the basis of perturbations of the stars' motions interpreted as due to gravitational influence of the companion. Whether or in what sense the dark companions are to be considered planets is not clear. Inferences as to size, radiation, and so on make them unlike any planets of our system and entirely unsuited for life.

Since much of the material in this chapter is recent, controversial, and somewhat outside my own field, it has seemed wise to document it in more detail than other chapters. The following explicit citations are made in the text:

Anders, E., and F. W. Fitch. "Search for Organized Elements in Carbonaceous Chondrites." *Science,* CXXXVIII (1962), pp. 1392–99).

Arrhenius, S. *Worlds in the Making* (New York: Harper and Brothers, 1908).

Barath, F. T., A. H. Barrett, J. Copeland, D. E. Jones, and A. E. Lilley. "Mariner II: Preliminary Reports on Measurements of Venus. Microwave radiometers." *Science,* CXXXIX (1963), pp. 908–9.

Becquerel, P. "La suspension de la vie des spores des bactéries et de moississures deséchées dans la vide vers le zéro absolu. Ses conséquences pour la dissémination et la conservation de la vie dans l'univers." *Comptes Rendus Acad. Sci. Paris,* CCXXXI (1950), pp. 1392–94.

Blum, H. F. "Negentropy and Living Systems." *Science,* CXXXIX (1963), p. 398.

Calvin, M. *Chemical Evolution.* (Eugene: University of Oregon Press, 1961).

———. "Communication: From Molecules to Mars." A.I.B.S. *Bulletin* (Oct. 1962), pp. 29–44.

Ehrensvärd, G. *Life: Origin and Development* (Chicago: University of Chicago Press, 1962).

Florkin, M., ed. *Some Aspects of the Origin of Life* (London: Pergamon Press, 1961).

Hawrylewicz, E., B. Gowdy, and R. Ehrlich. "Micro-organisms Under a Simulated Martian Environment." *Nature,* CXCIII (1962), p. 497.

Hess, H. H. (chairman), *et al. A review of space research.* Nat. Acad. Sci.–Nat. Res. Council, Pub. No. 1079, 1962. Includes a chapter of 23 pages on biology, ostensibly prepared by or expressing the views of 26 "principal participants," two or three of whom, only, are organismal biologists.

Hoyle, F. *Frontiers of Astronomy* (New York: Harper, 1955).

Jackson, F., and P. Moore. *Life in the Universe* (New York: Norton, 1962).

Keilin, D. "The Problem of Anabiosis or Latent Life: History and Current Concept." *Proc. Roy. Soc., B,* CL (1959), 149–91.

Kiess, C. C., S. Karrer, and H. K. Kiess. "A New Interpretation of Martian Phenomena." *Publ. Astron. Soc. Pacific,* LXXII (1960), pp. 256–67.

Levin, G. V., A. H. Heim, J. R. Clendenning, and M.-F. Thompson. "Gulliver—a Quest for Life on Mars." *Science,* CXXXVIII (1962), pp. 114–21.

Matthew, W. D. "Life on Other Worlds." *Science,* n.s., LIV (1921), pp. 239–41.

Miller, S. L. "Production of Some Organic Compounds Under Possible Primitive Earth Conditions." *Jour. Amer. Chem. Soc.,* LXXVII (1955), pp. 2351–61.

Nagy, B., W. G. Meinschein, and D. J. Hennessy. "Mass Spectroscopic Analyses of the Orgueil Meteorite: Evidence for Biogenic Hydrocarbons." *Ann. New York Acad. Sci.,* XCIII (1961), pp. 25–35.

———, G. Claus, and D. J. Hennessy. "Organic Particles Embedded in Minerals in the Orgueil and Ivuna Carbonaceous Chondrites." *Nature,* CXCIII (1962), pp. 1129–33.

Oparin, A. I. *The Origin of Life,* 3rd English (London: Oliver & Boyd, 1957).

———. "The Origin of Life on Earth." *Reports on the International Symposium of August, 1957, in Moscow* (Moscow: Pub. House Acad. Sci. USSR, 1959).

———, ed. *The Origin of Life on the Earth* (London: Pergamon Press, 1960).

Sagan, C. "Origin and Planetary Distribution of Life." *Radiation Res.,* XV (1961), pp. 174–92.

Shapley, H. *Of Stars and Men* (Boston: Beacon Press, 1958).

Simpson, G. G. *The Meaning of Evolution* (New Haven: Yale University Press, 1949).

———. "The History of Life." In *The Evolution of Life,* edited by S. Tax (Chicago: University of Chicago Press, 1960), pp. 117–80.

———. "Some Cosmic Aspects of Organic Evolution." In *Evolution und Hominisation,* edited by G. Kurth (Stuttgart: Gustav Fischer, 1962), pp. 6–20.

Sinton, W. M. "Further Evidence of Vegetation on Mars." *Science,* CXXX (1959), pp. 1234–37.

Chapter 14 Man's Evolutionary Future

I SPOKE on this subject at St. John's College, Annapolis, in April 1959, and later expanded and revised those lecture notes into an essay that was published in the *Zoologische Jahrbücher* (Syst. Bd. 88, pp. 125–134; 1960). Like the first published form of Chapter 12, this was included in a tribute to an honored friend and colleague in the study of evolution, in this instance Professor Bernhard Rensch. The present chapter reprints the essay without essential change.

There was, indeed, much temptation to add greatly to the text. This is a subject discussion of which could and perhaps should go on indefinitely. There is, in particular, much to add now that Th. Dobzhansky's splendid book *Mankind Evolving* (New Haven: Yale University Press, 1962) is available. But the very fact that the book is available makes expansion on its topics supererogatory. Furthermore, it is so richly documented that other citations are hardly needed here. Revision of my brief text would in any case have been mere expansion, because rightly or wrongly I still hold the opinions expressed in 1960.

It would have been more fun but less responsible to have taken an apocalyptic or a *Brave New World* view of man's evolutionary future. It is especially tempting (and others have yielded luxuriantly to the temptation) to weave fantasies on constructing people, or supermen, to order now that we have a faint glimmering of the genetic code. Science fictionists, some journalists, and even a few scientists anticipate unmade discoveries in such detail that real discoveries, if and when made, will come as anticlimaxes. T. M. Sonneborn has skillfully put this particular fantasy in its place, A.I.B.S. *Bulletin* (April 1963), pp. 22–26.